THE BASICS OF

PROJECT EVALUATION AND LESSONS LEARNED

THE BASICS OF

PROJECT EVALUATION AND LESSONS LEARNED

Willis H. Thomas, PMP

 CRC Press
Taylor & Francis Group
Boca Raton   London   New York

CRC Press is an imprint of the
Taylor & Francis Group, an **informa** business
A PRODUCTIVITY PRESS BOOK

CRC Press
Taylor & Francis Group
6000 Broken Sound Parkway NW, Suite 300
Boca Raton, FL 33487-2742

Printed in the United States of America on acid-free paper
Version Date: 20110504

International Standard Book Number: 978-1-4398-7246-8 (Paperback)

This book contains information obtained from authentic and highly regarded sources. Reasonable efforts have been made to publish reliable data and information, but the author and publisher cannot assume responsibility for the validity of all materials or the consequences of their use. The authors and publishers have attempted to trace the copyright holders of all material reproduced in this publication and apologize to copyright holders if permission to publish in this form has not been obtained. If any copyright material has not been acknowledged please write and let us know so we may rectify in any future reprint.

Library of Congress Cataloging-in-Publication Data

Thomas, Willis H.
   The basics of project evaluation and lessons learned / Willis H. Thomas.
      p. cm.
   Includes bibliographical references and index.
   ISBN 978-1-4398-7246-8
   1. Project management--Evaluation. I. Title.

HD69.P75T477 2012
658.4'04--dc23   2011017485

Visit the Taylor & Francis Web site at
http://www.taylorandfrancis.com

and the CRC Press Web site at
http://www.crcpress.com

Contents

Preface

When people come to understand Lessons Learned (LL) in the context of Project Evaluation (PE), they will soon come across the work of many renowned authors, such as Michael Scriven, Terry Williams, Bret Pettichord, Robert Brinkerhoff, Mark Kozak-Holland, and Michael Quinn Patton. This was my observation as I chose this subject matter for my dissertation at Western Michigan University. This topic was developmental in my academic life that began at the University of Wisconsin–Madison and professional life that began at Xerox.

This book is intended to be a practical guide to conducting project LL. It provides tools and techniques for active engagement. It is founded on the principles of conducting project evaluations as recommended by major governing bodies of project management, such as the Project Management Institute (PMI) and PRINCE2®. The concepts presented in this text are sound and utilized by the project management community, especially among those who are certified project managers.

This text simplifies and formalizes the methodology of conducting LL in projects. It purposes to fill a gap in content in the area of LL not fully delineated in the Project Management Body of Knowledge (PMBOK®) and Project in Controlled Environments version 2 (PRINCE2). My purpose in writing this book is to help organizations, large and small, to more effectively implement processes and systems to support effective LL. This text is supported by a Project Evaluation Resource Kit (PERK), which is found in CD format at the back of this book.

Willis H. Thomas, PhD, CPT, PMP

Acknowledgment

A special thanks to the many individuals and organizations that have contributed in some way to the development of this book. I would like to express my appreciation to *some of the many* professional associations identified that discuss Lessons Learned or Project Evaluation through their Web sites, articles, discussion boards, meetings, or other forms of communication:

ACP: Association of Career Professionals
AEA: American Evaluation Association
AERA: American Educational Research Association
AHA: American Hospital Association
AITP: Association of Information Technology Professionals
AMA: American Management Association
AMA: American Marketing Association
ANSI: American National Standards Institute
APA: American Psychological Association
APICS: Advancing Productivity Innovation for Competitive Success
ARL: Association of Research Libraries
ARMA: Association of Records Managers and Administrators
ASAE: American Society Association Executives
ASIS&T: American Society for Information Science and Technology
ASJA: American Society of Journalists and Authors
ASQ: American Society for Quality
ASTD: American Society for Training and Development
FDA: Food and Drug Administration
IAPPM: International Association of Project and Program Management
ICCP: Institute for Certification of Computing Professionals
ISA: Instrumentation Society of America
ISM: Institute for Supply Management
ISO: International Standards Organization
ISPI: International Society for Performance Improvement

ITAA: Information Technology Association of America
OSHA: Occupational Safety and Health Administration
PDA: Parenteral Drug Association
PMI: Project Management Institute
PRINCE: Projects in Controlled Environments
PRSA: Public Relations Society of America
SHRM: Society of Human Resource Management
SRA: Society of Research Administrators
TechAmerica
The Association of MBAs
The Conference Board
Toastmasters
United States Chamber of Commerce

The list goes on and on …

Introduction

Suppose you are planning to install a new computer network in your company. This network will have installations throughout the United States, Canada, Europe, Saudi Arabia, and South America. The computer network will be very complex because it interfaces with different systems, i.e., accounting, human resources, and customer service. It also utilizes multiple operating system platforms and configurations.

You are a Project Team Member (PTM) who has been assigned to implement a Windows®-based computer network. Upon completion of the computer network, you will be required to present an evaluation of the project through Lessons Learned (LL). Based upon the evaluation of this project, the company will determine the relative success or failure of the project as well as what could have been done differently. The company will use the LL to determine if and when it will expand the network to other current locations in Asia, Australia and Africa. So there is much communication from sponsors and key stakeholders to PTMs to ensure the Project Evaluation (PE) is properly conducted and LL are properly documented.

Regarding the background PTMs:

- Each person has different experience in Project Management (PM).
- Most of them have comparable levels of expertise in PM.
- Some are certified by the Project Management Institute (PMI) as Project Management Professionals (PMPs) and Certified Associates in Project Management (CAPMs).
- Others are certified by PRINCE2.
- Those who are not certified (i.e., by PMI or PRINCE2) have taken courses in PM.
- All have learned that evaluating a project is very important and know LL are valuable.

You have just received a call from the CEO of your company who has a big question for you regarding project cost. She wants to know how the budget is being managed and how close it is to the original estimate. She has requested a copy of the LL for the project to share it with the board of directors. You have searched far and wide for a guide on how to effectively conduct LL using sound Project Management and Evaluation (PM&E) principles. You have been able to locate a number of good books on LL and evaluating projects, but need a concise guide to distribute to PTMs. You also are discovering the wide range of perspectives from very simple to highly complex. Where do you go next?

For some organizations, LL is an informal process of discussing and recording project experiences during the closure phase. For other organizations, LL is a formal process that occurs at the end of each phase of a project.

Chapter 1

An Overview of Project Management

Key Learnings:

- Project, Program and Portfolio Management
- Process Groups
- Knowledge Areas
- Managing Stakeholders
- Organization Structures
- Organization Cultures
- Team Development Process
- Desired Competencies for Lessons Learned (LL) Project Team Members (PTMs)

What Is a Project?

A project is undertaken to address a requirement for a product, service, or result.

- A **product** is usually something physical, i.e., computer.
- A **service** is generally tangible, i.e., technical support response time.
- A **result** is typically directly associated with the item, i.e., based on this consumer research, the following forecast is recommended.

Why Project Management?

Project Management (PM) is the universally accepted standard for handling endeavors that are temporary, unique, and done for a specific purpose. PM engages good practices to support coordination, organization, and completion of projects from start to finish. PM involves the application of skills, knowledge, tools, and techniques. *Note*: A project should not be confused with ongoing operations. However, ongoing operations may involve projects.

With PM, you can meet defined objectives and satisfy stakeholders. Stakeholders are individuals who have a vested interest in the project. They can make or break the success of a project. Without PM, a multitude of problems can occur with respect to:

- Communications
- Cost
- Quality
- Resources
- Risk
- Scope
- Time

Project Management Process Groups

The Project Management Institute (PMI) addresses PM from two perspectives referred to as Process Groups (PGs) and Knowledge Areas (KAs) in the Project Management Body of Knowledge (PMBOK). The five PGs describe the work being performed in the project and are logically arranged PM activities that include:

1. Initiating
2. Planning
3. Executing
4. Monitoring/Controlling
5. Closing

Note: A Project Management Life Cycle (PMLC) typically involves four stages: initiate, plan, execute and close.

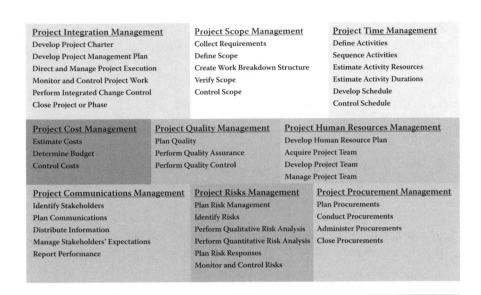

Figure 1.1   Project Management Knowledge Areas.

Project Management Knowledge Areas

The nine KAs depicted in Figure 1.1 define the required PM activities:

1. Communications
2. Cost
3. Human Resources
4. Integration
5. Procurement
6. Quality
7. Risk
8. Scope
9. Time

The five PGs and nine KAs are mapped in a table within the PMBOK to designate actions. For example, Project Risk is performed during the Planning and Monitoring/Controlling phase. *Note*: Please review the PMBOK to see the relationship denoted by intersection points between all PGs and KAs.

The objective of PM is to meet specified requirements as documented in the project plan. PM involves managing resources, cost, time, scope, quality, and risk to bring about successful completion of specific endeavors, which are temporary, unique, and for a specific purpose. The goal of PTMs is to satisfy customers by exceeding specified requirements in the project plan by demonstrating increased performance in aforementioned areas.

Boundaries of Project Management

PM is considered a:

- **Discipline:** It spans across industry types, involves a system of rules of conduct, or method of practice (PM code of ethics). It is interdisciplinary, intradisciplinary, and transdisciplinary.
- **Profession:** It is a recognized body of people in a learned occupation, typically requiring education, special skills, or certification.
- **Job Function:** A role or responsibility with clear expectations and measurable objectives.
- **Methodology:** A philosophical approach (involving tactical and strategic elements) that addresses how projects should be handled.
- **Process:** A series of steps or a system used to complete a requirement.
- **Activity:** A task or work item that is a component of a process.

Who Is a Project Manager?

A project manager (or project lead) is the person who is designated to manage PTMs. Because this is a lead position, the individual or group assigned should be authorized by the sponsor or key stakeholders. Qualifications of a project manager involve experience and frequently certification, i.e., Project Management Professionals (PMPs) from PMI or PRINCE2. Consult PMI's Competency Development Framework for Project Managers.

Programs versus Portfolios

Programs are interdependent, related projects for improved overall organizational alignment. For example, a company may be very diverse in its products or services and may categorize programs based upon similar categories, i.e., video games. This allows them to have more effective management of a similar group of projects. Program management combines knowledge, skills, and abilities of resources to define, plan, implement, and manage a series of projects. The benefits of program management include:

- Integration
- Harmonization
- Standardization

A portfolio is an appropriate mix of projects or programs based on key strategic attributes. Portfolio management seeks to determine the best combination of projects and programs, expressed in economic measures or strategic goals with consideration to real-world situations.

The portfolio represents the "package" of what stakeholders see as the organization's offerings (products, services, or results). Portfolio management addresses groups of projects or programs. Portfolio management provides a centralized view of an organization's project and program assets. For example, a portfolio may represent a consumer electronics division.

A project manager's role may encompass program or portfolio responsibilities. Respectively, a project manger would be considered a Tier 1, program manager Tier 2, and portfolio manager Tier 3. In this book, *project manager* represents the universal competency that also addresses program and portfolio manager responsibilities. PM represents the lowest common denominator and also addresses program and portfolio LL.

What Is Lessons Learned?

The term Lessons Learned (LL) refers to project learning, "ways of knowing" that have:

- Merit (quality)
- Worth (value) or
- Significance (importance)

LL may be integrated into a project or may be a project in itself. For example, a LL may concern the implementation of a computer network or in response to an initiative, such as recycling.

Who Is a Lessons Learned Project Team Member?

The success of LL as evaluation is heavily dependent upon the PTMs selected for the assignment. PTMs must be committed to Project Management & Evaluation (PM&E) and should appreciate the significant value of LL to the organization.

A PTM should be an employee (full or part time) or project consultant who is committed to:

- Project purpose, goals, and objectives
- Ethicality, honesty, and integrity
- Work ethic, productivity, and performance
- Effectiveness and efficiencies wherever possible

What Project Knowledge Is Required?

LL PTMs should have project knowledge in the following areas:

- Initiation, i.e., support stakeholders in a project idea and bringing it to reality.
- Planning, i.e., organize project details in a well-organized format.
- Executing, i.e., ensure project-related activities are carried out as planned.
- Monitoring/controlling, i.e., oversee evaluation, measurement, and research.
- Closing, i.e., end a project and release resources.
- Cost, i.e., know how to estimate costs and savings.
- Communications, i.e., demonstrate active listening skills.
- Human Resource, i.e., lead and manage a diverse group of people.
- Integration, i.e., see the big picture.
- Procurement, i.e., coordinate vendors and consultants.
- Quality, i.e., determine adherence to procedures and ensure compliance.
- Risk, i.e., identify gaps and areas of vulnerability.
- Scope, i.e., maintain objectives and administer change controls.
- Time, i.e., create schedules and balance resources.

What Competencies Are Essential?

Competency development, both behavioral (art) and technical (science), is a journey for PTMs. Required competencies for project evaluators will vary based upon the organization. Figure 1.2 is not intended to be exhaustive, but rather highlight some of the popular competencies that are required for those who will conduct lessons learned.

Team Development Process

While efforts should be made to support team development, PTMs will be subject to the typical five-stage process:

Evaluation	Professional Conduct	Tactical Planning	Meeting Facilitation
Measurement	Leadership	Strategic Planning	Active Listening
Research	Data Management	Continuity Planning	Oral Communication
Systems Thinking	Team Building	Succession Planning	Non-verbal Communication
Critical Thinking	Performance Management	Project Planning	Interpersonal Communication
Creative Thinking	Relationship Management	Emotional Intelligence	Written Communication
Problem Solving	Delegating	Self-Control	Computer Software Technology Use

Figure 1.2   Example of Competencies

1. Forming: PTMs are introduced
   a. PTMs are driven by a desire to be accepted by the other PTMs.
   b. PTMs will likely avoid conflict and be reserved in who they are as a person.
   c. During this stage, it can be helpful to introduce a personality profile and skills summary to uncover personal traits and determine work experience.
2. Storming: The PT transitions to initial confrontation
   a. Team members open up to each other and may confront each other's ideas.
   b. Some individuals may feel in competition regarding their expertise or position.
   c. People will seek to find their place and understand expectations.
3. Norming: The team establishes parameters for the project
   a. The team agrees upon the direction as specified by the project lead.
   b. A teaming agreement should be constructed to outline PTM code of conduct.
   c. Performance expectations, roles, and responsibilities should be outlined.

	Organizational Structure				
		Matrix			
	Functional	Weak	Balanced	Strong	Projectized
Project Manager's Authority	Limited	Low	Low to Moderate	Moderate to High	High
Resource Availability	Low	Limited	Low to Moderate	Moderate to High	High
Project Budget	Functional Manager	Functional Manager	Shared	Project Manager	Project Manager
Utilization	Project Manager and Staff Part-time	Project Manager and Staff Part-time	Project Manager Full-time, Staff Part-time	Project Manager and Staff Full-time	Project Manager and Staff Full-time

Figure 1.3   Project Management interface to organizational structure.

4. Performing: The team engages in the project
   a. Relationships are at a comfortable working level.
   b. Expectations are understood and adhered to.
5. Adjourning: The team transitions
   a. The team revisits their experience as a LL.
   b. There are feelings of accomplishment.
   c. There are feelings of sadness due to project closure.

Organization Structures

The type of organization structure (environment) is a consideration with regard to how projects are evaluated and how LL are addressed. The type of organizational structure not only impacts reporting relationships, but also can influence the number of resources allocated to projects. Figure 1.3 depicts organizational structure.

Organizational dynamics for a functional organization is different than a projectized organization. For example, in a functional organization structure, employees are grouped hierarchically, managed through clear lines of authority, and report to a person responsible for that area. A projectized organization resembles more of a homoarchy or heterarchy in design. Independent of the type of organization structure, there is a need to coordinate project activities. To support the effective and efficient management of projects, an organization may choose to create a PM department, commonly referred to as a PMO (Project Management Office).

Organization Cultures

There are basically three types of organization culture:

1. **Bureaucratic:** Governed by politics, i.e., local, state, and federal agencies. Examples include police departments, schools, and social service agencies.
2. **Rules-based:** Driven by policies and procedures, i.e., regulated industries. Examples include pharmaceutical and healthcare.
3. **Learning Organization:** Prompted by continuous improvement, i.e., marketplace demands. Examples include information technology and beverage companies.

Who Are Project Stakeholders?

A project stakeholder is any person with an interest in the project, program or portfolio initiative:

- Employees
- Vendors: consultants, contractors, suppliers
- Customers
- Regulators

Understanding who they are is critical to project success and involves:

- Knowing who they are and what they will do
- Identifying their personality style
- Classifying them into groups
- Engaging them actively in the process as appropriate

A Key Project Stakeholder (KPS) is an individual whose involvement is integral to the success of the project and thereby LL. For example:

- Sponsor: members of senior management
- Subject Matter Expert (SME): content provider, reviewer or approver

- Project Team Member: Project manager, project lead, project coordinator, etc.
- General Lessons Learned Service Provider: (GLLSP) entity that provides project evaluation related support, i.e., benchmarking data or LL workshop facilitation

Working with Project Stakeholders

Stakeholder management is a proactive process that requires patience and attention to detail. Using Table 1.1, list KPS involvement with the project and assign a PTM to each KPS.

Working with stakeholders can be challenging. PTMs must understand human behavior and be committed to maintaining good working relationships. Below are a few good practices worth considering.

Relationship building includes:

- Clarifying expectations of who is doing what: when, how, why and where
- Identifying the necessary resources to address project-related activities
- Supporting ongoing awareness and understanding of project issues

Managing stakeholders involves:

- Communication: distributing information and engaging in discussions
- Collaboration: working to gain agreement and determine paths forward
- Coordination: addressing change and new ways of doing things

Table 1.1 Project Stakeholder Register

Project Stakeholder Register				
Name	Title	Level	Function	PTM Assigned
Sara N. Charge	CFO	Senior Management	Sponsor	John Teamers
John Brain	Vice President	Senior Management	SME	Sally Friends
Function: Sponsor, Subject Matter Expert (SME), Supplier (Vendor, Contractor, Consultant)				

Leading stakeholders in the preferred direction requires:

- Persuasion: influencing
- Negotiation: reaching agreement
- Conflict resolution: reacting to a difficult situation with an agreed upon solution

Use the STYLE Framework to Discover Stakeholder Style

Discovering the personality traits of stakeholders is core, common and critical to interpersonal communication. Knowing what works and what does not work with people's styles is essential to gaining agreement. Using a STYLE chart has been proven to be an effective way to ensure consistent results in working with people.

People have a tendency to be more informal at home then at work. So it may be impractical to use one chart to generalize populations in every environment. Moreover, this framework is a high-level perspective that serves as a starting point for personality assessment. This framework is designed to have application to the job or structured setting. Although every stakeholder personality is unique within the workplace, most can be grouped into one of four areas as illustrated in Figure 1.4.

Use the TYPE Grid to Classify Stakeholders

After determining the STYLE(S) for each stakeholder, place them in the Stakeholder Grid below (Figure 1.5).

Note: Grid-lock represents barriers in communication that may occur when personality conflicts arise. This chart is instrumental in helping to determine approaches to work with stakeholders.

Resolving Stakeholder Neutral Opinions

Stakeholders need to clear in their position on issues related to projects. Remaining on the "fence" is an unacceptable position. Neutrality is the "0" point on a response scale that does not provide any indication of their thoughts or

Strong **T**raits **Y**ou **L**ikely **E**xhibit	
• Methodological • Factual • Analytical	• Determined • Dominant • Decisive
Logic & Systems	**Tasks & Results**
Harmony & Relationships	**Show & Tell**
• Supportive • Empathetic • Cooperative	• Verbal • Impulsive • Convincing
Framework for Understanding Behavior	

Figure 1.4 Stakeholder STYLE Framework

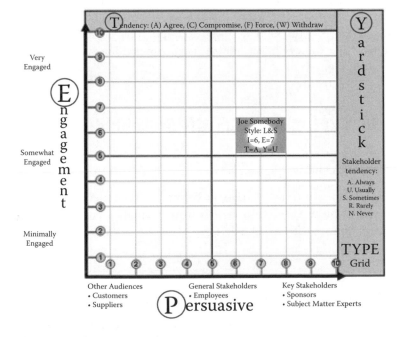

Figure 1.5 Stakeholder Type Grid

potential actions. In the context of LL, it is important to resolve neutral thinking because it can not be interpreted. Neutrality may display the following behaviors:

- **Nothing was done right:** negativity to a degree that overshadows any reasonable contribution or communication
- **Non-participation:** silence verbally or non-verbally or not in attendance
- **No opinion whatsoever:** can't or won't share any perspectives on any events, but is present
- **Absent presence:** unengaged and involved in other activities during discussion of LL, i.e., writing emails, texting, etc.
- **No opinion whatsoever:** no considerations (meaningful thoughts) or issues (concerns) expressed
- **Indifference/Don't care:** have no expressed interest one way or the other
- **No identified issues:** not perceiving anything could be a concern, problem, etc.
- **Everything was done right:** no critical thinking to review gaps or areas not meeting expectation

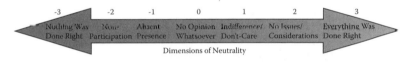

Figure 1.6   Range of Neutral Response

Neutral thinking can be resolved by moving towards a productive thought process:

- **Abstract thinking:** brainstorming and reverse brainstorming attempt to inspire freedom of thought to capture ideas of value.
- **Creative thinking:** is an inventive process that draws upon new concepts and ideas to improve things in the future. Creative thinking is optimistic, hopeful and best case oriented. Creative thinking is a key attribute of LL and embellishes "What If" scenarios to enable higher-order thoughts.
- **Critical thinking:** involves a complex set of cognitive skills and logical principles to reach a determination of the preferred solution. Critical thinking is neutral, negative or questioning in its problem-solving nature because its purpose is to identify gaps, alternatives or solutions.

Systems thinking: reviews inter-relationships and influencing factors to determine the steps, sequences, or stages something goes through during it typical life cycle.

Project Management Office

The purpose of a PMO is to centralize project administrative, technical, and management support functions. The PMO can support the effective alignment of project resources (i.e., people and systems) to project stakeholders. This will improve project communication resulting in many benefits to the organization. A Project Management Center of Excellence (PMCOE) refers to best practices held by the PMO. PMO also may refer to a Program Management Office or Portfolio Management Office. Activities performed by a PMO include:

- PM budgets, i.e., contingency reserves
- PM resources, i.e., contractor selection
- PM templates, i.e., checklists and forms
- PM training, i.e., PMP® certification
- PM methodology, i.e., PMI PMBOK or PRINCE2
- PM best practices, i.e., benchmarking data
- Project Management Information System (PMIS) i.e., MS Project Reports
- PM communications, i.e., status reports
- PM documentation, i.e., LL

Lessons That Apply to This Chapter

1. Projects are endeavors that are temporary, unique and done for a specific purpose.
2. There should be an adequate investment of time, money, and energy in PM&E training programs to ensure the desired level of competency of PTMs.
3. A common language for PM&E must be established, which can be done through a Web-based online glossary. Using common terms and definitions is essential to promoting effective communication among PTMs.
4. PM processes involve initiating, planning, executing, monitoring/controlling, and closing. These processes should be a part of all projects.
5. A PLC includes basically four steps: initiating, planning, executing, and closing. The PLC is iterative in reality, i.e., back and forth engagement between planning and executing activities before project closure.
6. Project knowledge areas are actions the PTMs perform during the PLC, i.e., communications, cost, human resources, integration, procurement, quality, risk, scope, and time.
7. A LL PTM is an individual who is capable of performing PM&E-related activities. This person must be invested in performing quality work.
8. The typical team development process (forming–storming–norming–performing–adjourning) is usually improved when this model is described to

PTMs at the formation of the team and processes are put in place to support relationship building.

9. A PMO can be an excellent resource provided that it is staffed appropriately.
10. Organizational structure and culture are important to how PM&E are handled.

Suggested Reading

Altschuld, J. 1999. The certification of evaluators: Highlights from a report. *American Journal of Evaluation* 20, 481–493.

American Evaluation Association. 2007. *Guiding principles for evaluators.* Fairhaven, MA: AEA.

Crowe, A. 2006. *Alpha project managers: What the top 2% know that everyone else does not.* Kennesaw, GA: Velociteach, Inc.

Davidson, J. 2005. *Evaluation methodology basics: The nuts and bolts of sound evaluation.* Thousand Oaks, CA: Sage Publications, Inc.

Dinsmore, P., and J. Cabanis-Brewin. 2010. *The AMA handbook of project management,* 2nd ed. New York: Amacom.

Fabac, J. 2006. Project management for systematic training. *Advances in Developing Human Resources* 8, 540–547.

Heldman, K. 2005. *Project management professional study guide,* 4th ed. Hoboken, NJ: Wiley Publishing, Inc.

Kendrick, T. 2004. *The project management toolkit: 100 tips and techniques for getting the job done right.* Saranac Lake, NY: Amacom.

King, J., L. Stevahn, G. Ghere, and J. Minnema. 2001. Toward a taxonomy of essential evaluator competencies. *American Journal of Evaluation* 22, 229–247.

Kotnour, T. 2000. Leadership mechanisms for enabling leaning within project teams. Online from: www.alba.edu.gr/OKLC2002/Proceedings/ pdf_files/ID340.pdf (accessed December 12, 2007).

Llewellyn, R. 2006. PRINCE2 vs. PMP. Online from http://manage.wordpress. com/2006/11/24/prince2-vs-pmp/ (accessed December 21, 2007).

Marsh, D. 1996. Project management and PRINCE. *Health Informatics* 2, 21–27.

Mathison, S., ed. 2005. *Encyclopedia of evaluation.* Thousand Oaks, CA: Sage Publications, Inc.

Mulcahy, R. 2005. *PMP exam prep: Rita's course in a book for passing the PMP exam,* 5th ed. Minneapolis, MN: RMC Publications, Inc.

Project Management Institute. 2002. *Project manager competency development framework.* Newtown Square, PA: Project Management Institute.

Project Management Institute. 2003. *Organizational project management maturity model (OPM3): Knowledge foundation.* Newtown Square, PA: PMI.

Project Management Institute. 2007. Source: *Project management institute code of ethics and professional conduct.* Newtown Square, PA: PMI.

Project Management Institute. 2009. *A guide to the project management body of knowledge,* 4th ed. Newtown Square, PA: PMI.

Raupp, M., and F. Kolb. 1990. *Evaluation management handbook.* Andover, MA: Network, Inc.

Rosas, S. 2006. A methodological perspective on evaluator ethics. *American Journal of Evaluation* 27, 98–103.

Sartorius, R. 1991. The logical framework approach to project design and management. *American Journal of Evaluation* 12, 139–147.

Scriven, M. 1991. *Evaluation thesaurus,* 4th ed. Newbury Park, CA: Sage Publications, Inc.

Scriven, M. 1996. Types of evaluation and types of evaluator. *American Journal of Evaluation* 17, 151–161.

Scriven, M. 2007. Key evaluation checklist. Online from http://www.wmich.edu/evalctr./checklists /checklistsmenu.htm (accessed April 25, 2007).

Smith, M. 1999. Should AEA begin a process for restricting membership in the profession of evaluation? *American Journal of Evaluation* 22, 281–300.

Sneed, J., V. Vivian, and A. D'Costa. 1987. Work experience as a predictor of performance: A validation study. *Evaluation of Health Professionals* 10, 42–57.

Stevahn, L., J. King, G. Ghere, and J. Minnema. 2005. Establishing essential competencies for program evaluators. *American Journal of Evaluation* 26, 43–59.

Sundstrom, E., and D. Gray. 2006. *Evaluating the evaluator.* National Science Foundation 2006 Annual Meeting. Online from http://72.14.205.104/ search?q=cache:NLb pOJqJuBkJ:www.ncsu.edu/iucrc/Jan%2706/Evaluate%2520the%2520Evaluator. ppt+evaluating+the+evaluator&hl=en&ct=clnk&cd=5&gl=us (accessed December 14, 2007).

Yang, H., and J. Shen. 2006. When is an external evaluator no longer external? Reflections on some ethical issues. *American Journal of Evaluation* 27, 378–382.

Zigon, J. 1997. Team performance measurement: A process for creating team performance standards. *Compensation Benefits Review* 29, 38–47.

Zwerman, B., J. Thomas, S. Haydt, and T. Williams. 2004. *Professionalization of project management: Exploring the past to map the future.* Philadelphia, PA: Project Management Institute.

Chapter 2

Foundations of Evaluation

Key Learnings:

- Origins of evaluation
- Potential complexity of project evaluation
- Evaluation approaches and models
- Classifications of evaluation
- Component evaluation
- Outcome evaluation
- Logic models
- Summative and formative evaluation
- Descriptive and evaluative knowledge
- Tacit and explicit knowledge
- DIKUD continuum

Origins of Evaluation

Evaluation is the determination of Merit (quality), Worth (value), or Significance (importance). Since the dawn of man, evaluation has been in use. People have used evaluation to prove and make judgments. People have used Lessons Learned (LL) as a tool of evaluation and a means of storytelling. These stories have been used to:

- Document results or determine accountability (summative)
- Improve things for the future or determine other options (formative)

Lessons Learned in Project Management

LL involves engaging evaluation in projects at a tactical and strategic level. The context of LL can be seen throughout history, even in many ancient writings, such as the Dead Sea Scrolls. Over the past four hundred years, the concept of Project Evaluation (PE) can be credited to pioneers such as Francis Bacon. In 1620, Bacon was conducting evaluations and involved with projects. The scientific method he designed was **hypothesis–experiment–evaluation** and **plan–do–check**. Three hundred years later, in the early 1900s, Walter Andrew Shewhart extended the **plan–do–check** to the **plan–do–check–act**. In the 1920s, Gantt charts were introduced by Henry Laurence Gantt to visually display scheduled and actual progress in parts of projects. By the late 1950s, **Critical Path Method** (CPM) was introduced by DuPont and Remington Rand Univac to coordinate complex plant maintenance projects (Fabac, 2006).

By the 1960s, the plan-do–check–act model became popularized by William Edwards Deming as the **plan–do–study–act**, which provides a continuous improvement focus within the context of quality, performance, and managing projects (Kotnour, 2000). In the Plan phase, the Project Team (PT) determines the nature of the problem and constructs a plan. The **Plan** phase is a series of expectations about the necessary steps to take and the anticipated results. In the **Do** phase, the PT implements the plan. Implementation creates a series of results regarding the anticipated and unanticipated actions taken and associated performance, such as cost, schedule, or technical performance. These findings are used to comprehend project status and to move the project forward. In the **Study** phase, the PT reflects on the applicable plans and results to determine good and bad situations. The output of the study step is the **Act** Phase, which encompasses evaluating projects and lessons.

A connection occurred between LL and PM in the mid-1960s when Middleton (1967) linked LL to PM by coining the term *Lessons Learned*. Middleton stated that LL on one project should be communicated to other projects. In 1969, the Project Management Institute (PMI) was formed, which purposed to increase professionalism in PM and standardize PM processes.

During this period, in the mid-1960s, Daniel Stufflebeam developed the CIPP model (Context, Inputs, Processes, and Products). It contains four stages of assessment: Context (requirements), Input (alternatives), a Process (implementation plans), and a Product (outcomes). The CIPP model emphasizes evaluation's main purpose is to improve (formative) as compared to other evaluation models, which purpose is to prove.

During the late 1960s, Michael Scriven coined the terms summative and formative evaluation.

■ Summative evaluation is conducted to determine the results, accountability or achievement of objectives in relation to the resources deployed. Resources may include people, systems, and materials. It can be described as looking through the rearview mirror to see the results of what passed by.

■ Formative evaluation is commonly referred to as developmental evaluation or implementation evaluation and purposes to improve the project. It can be described as looking through the windshield to see what is coming and making adjustments.

Scriven also introduced meta-evaluation during the late 1960s. It purposes to:

■ Evaluate the evaluation—using the same or different criteria

■ Evaluate the evaluator—quality of work performed and adherence to criteria

So, it could be argued that Project Management & Evaluation (PM&E) in the context of LL began to springboard during the late 1960s. This revitalized introduction of PM&E is the result of project managers, evaluation theorists, and practitioners, such as Cronbach, Guba, Stufflebeam, Scriven, and Middleton. Since then, new roots of evaluation have been growing, engaging a growing number of people who use a wide variety of tools and techniques to support PM&E.

Alkin (2004) and Mathison (2005) are definitive sources on tracing new developments in evaluation. During the early 1980s, there would be a formal development of LL for the U.S. army that recognized there was no formal procedure to capture the war-fighting lessons coming from the training center in the Mojave Desert, despite the significant investment in the National Training Center (NTC). Therefore, the U.S. army created the Center for Army Lessons Learned (CALL) in 1985 at Fort Leavenworth, Kansas. During this same period in the early 1980s, technology began to play a major role to support PM&E. The introduction of the IBM Personal Computer (PC) in 1981 is perhaps the most significant technological achievement of the twentieth century. The PC was further enhanced by the public introduction of the Internet in 1992. Moreover, the portable document file (PDF) created by Adobe® in 1993 made it possible to move documents to an electronic format.

Since the early 1990s, the Internet continues to be enhanced. Mobile computing devices stream the airways. Personal digital assistants (PDAs) make it possible to move personal computing power from the desktop and laptop to handheld. What this means for evaluators is there is now a tool that can provide for improved management of electronic files. Digital diaries can be used to record and track evaluations as lessons, which can be stored real time in a repository.

Classifications of Evaluation

There are primarily six ways to categorize evaluation:

1. Project Evaluation
2. Product Evaluation
3. Program Evaluation
4. Policy Evaluation
5. Process Evaluation
6. Performance Evaluation

Is Project Evaluation Complex?

PE can be complicated or noncomplex depending upon a multitude of factors. At the complex end, the Analytic Hierarchy Process (AHP) for the evaluation of project termination or continuation is based on a benchmarking method and utilizes statistical models to derive conclusions. LL, comparatively speaking, is a straightforward process, based on the three dimensions (e.g., done right, done wrong, done differently, or variations thereof). While most every project can benefit from LL, it is not the only approach to evaluating a project. The key to successful PE is to determine the value that evaluation brings and what stakeholders desire.

What Is Component Evaluation?

Component evaluation (CE) is an assessment of each aspect of the evaluand (project), and the sum of the whole consisting of a report on the merit (relative quality based on criteria) of each item. Components, including each of the five PM process groups (Initiation through Closure) and nine PM knowledge areas (Communications through Time), are analyzed.

What Is Outcome Evaluation?

Outcome evaluation (OE) asks whether a project has met its goals. OE purposes to establish the impacts of the project and measure changes in outcomes of the project. Therefore, a relationship exists between LL and OE, which can be depicted through a logic model (Figure 2.1).

Lessons Learned Is Project Learning

The term *Lessons Learned* (LL) refers to project learning, "ways of knowing" that have merit (quality), worth (value), or significance (importance), which has

Lessons Learned Project Management Logic Model

| Project Charter | Current Situation (Need) A description of the issue or concern the project seeks to address |

| Goal (Objective) The purpose of the project, including vision and mission of the project team |

| Assumptions (Rationale) Why project activities will produce results considering factors already in place. | Resources • Employees • Companies • Contractors • Suppliers • Regulators • Equipment • Systems • Facilities • Supplies • Material | Activities Actions project team takes to achieve results. PLC: Initiating. Planning, Executing, Closing with Monitoring / Controlling | Outputs (Deliverables) Direct and Indirect Products, Services Results resulting from project |
| Constraints (Barriers) What factors could impact the ability for the project to be successful, i.e., cost, time, scope, risk, etc. | ...that will be involved with project objectives | Communications, Cost, Human Resources, Integration, Procurement, Quality, Risk, Scope and Time | Outcomes (Impact) Anticipated results due to project success or failure, completion, termination, or reschedule. Lessons Learned Summary Formative or summative depending upon context of project. |

Figure 2.1   Lessons Learned Project Management Logic Model.

occurred through Reference, Observation, Participation, or Experience (ROPE) (Figure 2.2):

- Reference: No involvement, i.e., communities of practice
- Observation: Indirect involvement, i.e., hearing or seeing
- Participation: Direct involvement in an activity
- Experience: Tangible and intangible series of events that results in ways of knowing

In LL, there is valuable learning through experience that can be captured and documented. LL, in the context of a PM, reviews a wide range of topics. For example, there may be a project undertaken to bring public awareness to forest fires. In this example, the deliverable is a result, which is to prevent forest fires. Information may include:

- How should liquids or solids be cooked when camping?
- What must be done to ensure increased care with flammable liquids?
- When are the safest time periods for hiking?

Reflecting on the example above, policies or procedures may be established to include:

- Flyers for campers visiting natural parks
- Signage posted throughout walkways and paths
- Periodic supervision to ensure the hiker safety is enforced

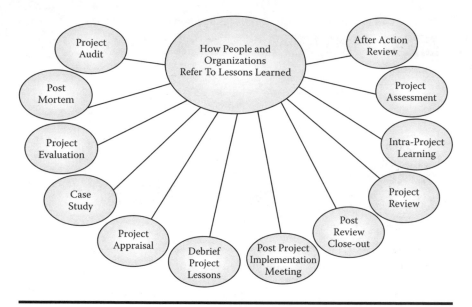

Figure 2.2 Common names for Lessons Learned.

How Can Lessons Learned Be Used?

There are many things that LL can be used for. Some examples include:

- Performance Evaluation (summative)
- Regulatory Audit (summative)
- Earned Value Analysis (summative)
- Business Continuity Planning (formative)
- Strategic Planning (formative)
- Succession Planning (formative)

If an organization seeks to assess and analyze learned experiences, LL can be both of the following:

- Effective: Preferred method to collect and document project component elements
- Efficient: Lower cost way of conducting an evaluation using available resources

Is Intraproject Learning Feasible?

Yes! Intraproject learning also known as (AKA) LL is feasible. Kotnour (2000) defines intraproject learning as the creation and sharing of knowledge within a project. Intraproject learning focuses on activities within a project, and using LL to resolve problems during a project. Reeve and Peerbhoy (2007) explain that a key aim should be to learn from the evaluation process and understand how evaluation contributes to future learning for all stakeholders on the project.

Similarly, Ladika (2008) states that project managers should always be prepared to recognize a LL opportunity. This may come when the team is complaining or it may come at the end of a project when the team is recapping areas for improvement. Ladika identifies a PE as a necessary tool to ensure this information is captured.

What Investment Is Required to Conduct Lessons Learned?

An investment in the context of conducting LL represents the commitment of primarily time, energy, and resources (sometimes money). The monetary requirement for conducting LL is usually reasonable. It may include the cost of facilities, facilitator, and supplies. Staff time should be noted in the sense of perceived lost productivity due to meeting time. *Note*: These time observations should only be used to encourage making the time investment productive.

Determining the value of LL can be done by:

- Quantitatively: Looking at
  - Return on investment
  - Payback period
  - Internal rate of return

- Qualitatively: Reviewing
  - Increases in customer satisfaction
  - Improvements in workflow, i.e., documents
- Influences in quality deliverables

Using these methods together in unison results in a mixed-method approach.

What Is Appreciative Inquiry?

Appreciative Inquiry (AI) is one method to justify the investment in LL. AI is an over-arching concept that investigates what an organization is doing well so that it can be replicated. It is an organizational development process that finds the answer to important questions. It utilizes LL to engage individuals in accepting accountability and supporting process improvement.

Descriptive versus Evaluative Knowledge

Descriptive knowledge (DK) is classified as declarative knowledge or propositional knowledge. DK is expressed in declarative sentences or indicative propositions. DK involves generalized ways of knowing that describes the physical attributes or characteristics of the evaluand (item under evaluation). DK is broad and all-encompassing, hence, there is lots of knowledge classified as descriptive. For example, "the car is red" is descriptive knowledge. The car being red doesn't address merit, worth, or significance. For the purpose of LL, therefore, DK alone does not provide the level of depth required to prove or render judgments. Thus, collecting only DK alone from a project is insufficient to arrive at evaluative conclusions. In order for knowledge to be valuable, Project Team Members (PTMs) must move beyond the collection of just DK.

On the other hand, Evaluative Knowledge (EK) is processed (synthesized) knowledge that addresses merit (quality), worth (value), or significance (importance) of the evaluand. For example, the red Ferrari costs more than $100,000, is best in class, and is fast to drive. *LL is EK.* Consider this: How do you know? I know that I know! This simply is not good enough! The term *Knowledge Management* (KM) has been popularized by management theorists, such as Peter Drucker, Paul Strassmann, and Peter Senge. The concept of EK has more recently been introduced by evaluation theorists, such as Ray Rist, Nicoletta Stame, and Jody Kusek, and can be viewed as another branch in the growth of KM that has direct application to LL. EK represents ways of knowing that involves a determination of merit (quality), worth (value), or significance (importance). This is why LL more closely represents EK, not just general or ways of knowing (DK). The need for EK in a performance-based system is not so much to prove, but to improve (Rist and Stame, 2006).

Tacit, Explicit, and Embedded Knowledge

Knowledge is a "fluid mix of framed experiences, values, contextual information, expert insight, and intuition that provides an environment for evaluating and incorporating new experiences and information" (Tiwana, 2002).

Knowledge is a system of connections between facts and ideas (Tywoniak, 2007). As Nelson (1979) explains, the "knowledge creation loop" involves creation, diffusion, utilization, and evaluation. A growing trend amongst knowledge management theorists is to classify knowledge in two dimensions: explicit and tacit (Tiwana, 2002). Explicit knowledge is skill-based knowledge. It is knowledge that is relied upon qualification, certification, or other form of verification. How to drive a car or use a software program are examples of explicit knowledge. We use the terms *smarts* or *IQ* to address an individual's ability to maximize explicit knowledge.

Tacit knowledge is interpersonal, a form of personal know-how (Tywoniak, 2007). Brockmann and Anthony (2002) state tacit knowledge resides just outside of our active conscious, which stores data currently being perceived by our senses. Professional singing is an example of tacit knowledge. This innate ability to master a range of sounds represents personal know-how.

Tacit knowledge may be considered "natural talent" by some critics. However, it may actually be more complex when one considers the implications of knowing. In other words, tacit knowledge involves knowing how to sing, not necessarily the talent that makes one sing. For example, a musician may understand all of the science behind the notes, can play an instrument, but not be a singer. Other nonmusicians may not understand the science behind the music, which has vast implications for how music is created. Where explicit knowledge and tacit knowledge intersect is the optimum solution for growth and development.

Embedded knowledge refers to the decisions and understanding that is linked to policies, practices, processes, and procedures as well as products, services, culture, and structure. Embedded knowledge can be formal or informal, written or oral in nature. An organization's code of conduct, ethics, guiding principles, mission, vision and values, and related knowledge assets represent embedded knowledge.

Determining Competence and Defining Competencies

There are four variations for how competence is perceived:

1. **Conscious competence:** The person must concentrate to demonstrate the skill, but can usually perform it at will.
2. **Conscious incompetence:** The person is aware of a deficiency in the skill and must put measures in place to improve to become consciously competent.
3. **Unconscious competence:** The person does not need to concentrate to demonstrate the skill because it has become second nature. It is a natural reflex.

4. Unconscious incompetence: The person is not aware of his/her deficiency in the skill and, therefore, may not invest the level of effort to become proficient.

There are two classifications of competencies: technical and behavioral.

1. Technical: Skills, Talent, Aptitude, Capabilities, and Knowledge (STACK) that are required to perform the work required.
2. Behavioral: Personality, Intuition, Language, and Emotions (PILE) that are necessary to enable acceptance of the person to perform the work required. *Note*: Language includes any characteristic that impacts communication, i.e., body language (nonverbal).

What Is the DIKUD Continuum?

The Data–Information–Knowledge–Understanding–Decision (DIKUD) continuum is a way to look at the relationship involving data management to decision making. In this model:

■ Data serves as an input to Information.
■ Information becomes the source for Knowledge.
■ Knowledge supports the development of Understanding.
■ Understanding cognitive process results in a Decision.

A similar theory in the form of a reverse triangle was introduced by the U.S. army as *Information Theory as Foundation for Military Operations in the 21st Century* (Sparling, 2002). In their design (bottom to top), Data (processing), Information (cognition), Knowledge (judgment), and Understanding was the flow. The model, while technically sound, provided no horizontal categorization to support more definition for application to LL.

The DIKUD in Figure 2.3 represents an adapted model for use with LL. It considers five steps and provides horizontal categories.

References

Brockmann, E., and W. Anthony. 2002. Tacit knowledge and strategic decision making. *Group Organization Management* 27, 436–455.
Fabac, J. 2006. Project management for systematic training. *Advances in Developing Human Resources* 8, 540–547.
Kotnour, T. 2000. Leadership mechanisms for enabling learning within project teams. Online from: www.alba.edu.gr/OKLC2002/Proceedings/ pdf_files/ID340.pdf (accessed December 12, 2007).

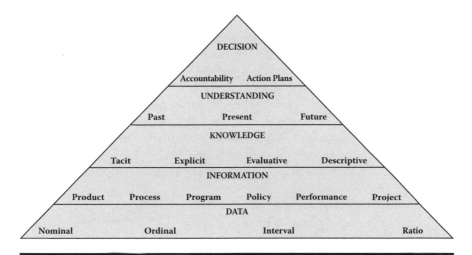

Figure 2.3   The DIKUD continuum represents an adapted model for use with LL.

Ladika, S. 2008. By focusing on lessons learned, project managers can avoid repeating the same old mistakes. *PM Network*, February.

Middleton, C. 1967. How to set up a project organization. *Harvard Business Review* March–April, 73–82.

Nelson. 1979.

Reeve, J., and D. Peerbhoy. 2007. Evaluating the evaluation: Understanding the utility and limitations of evaluation as a tool for organizational learning. *Health Education Journal* 66, 120–131.

Rist, R., and N. Stame. 2006. *From studies to streams: Managing evaluative systems.* New Brunswick, NJ: Transaction Publishers.

Sparling. 2002.

Tiwana. 2002.

Tywoniak. 2007.

Lessons That Apply to This Chapter

1. LL is a form of PE because it involves the determination of merit (quality), worth (value), or significance (importance).
2. There is a natural synergy between evaluation and LL.
3. There are different ways to conduct a PE. LL, comparatively speaking, can be a more effective and efficient way depending on the situation.
4. While one could associate the concept of LL back to 1620 with Francis Bacon, more formal developments occurred during the 1960s with support from theorists, such as Scriven, Cronbach, Stufflebeam, and Middleton.

5. Organizations may refer to the term LL in different ways.
6. The PM knowledge areas have individual characteristics that are referenced in the Project Management Body of Knowledge (PMBOK). These individual items are essential to gaining a comprehensive perspective on activities involved in the Project Life Cycle (PLC).
7. LL can be conducted summatively to determine impact and/or formatively to address improvements.
8. Evaluative knowledge is the framework that attempts to provide the necessary level of depth to make evaluative judgments.
9. A person should have both technical and behavioral competency development plans.
10. LL can be used for many applications and is situation-specific depending on the needs of the organization.

Suggested Reading

Brown, R., and C. Reed. 2002. An integral approach to evaluating outcome evaluation training. *American Journal of Evaluation* 23, 1–17.
Cresswell, J., and V. Clark. 2007. *Designing and conducting mixed methods research.* Thousand Oaks, CA: Sage Publications, Inc.
Davidson, J. 2005. *Evaluation methodology basics: The nuts and bolts of sound evaluation.* Thousand Oaks, CA: Sage Publications, Inc.
Gray, P. 1989. Microcomputers in evaluation: People, activity, document, and project organizers. *American Journal of Evaluation* 10, 36–43.
MacMaster, G. 2000. Can we learn from project histories? *PM Network*, July. Philadelphia, PA: Project Management Institute.
Preskill, H., and T. Catsambas. 2006. *Reframing evaluation through appreciative inquiry.* Thousand Oaks, CA: Sage Publications, Inc.
Preskill, H., and R. Torres. 1999. *Evaluative inquiry for learning in organizations.* Thousand Oaks, CA: Sage Publications, Inc.
Reider, R. 2000. *Benchmarking strategies: A tool for profit improvement.* Hoboken, NJ: John Wiley & Sons.
Renger, R., and A. Titcomb. 2002. A three-step approach to teaching logic models. *American Journal of Evaluation* 23, 493–503.
Rodriguez-Campos, L. 2005. *Collaborative evaluations: A step-by-step model for the evaluator.* Tamarac, FL: Llumina Press.
Scriven, M. 1991. *Evaluation thesaurus*, 4th ed. Newbury Park, CA: Sage Publications, Inc.
Shaw, I., J. Greene, and M. Mark. 2006. *The Sage handbook of evaluation.* Thousand Oaks, CA: Sage Publications, Inc.
Torraco, R. 2000. A theory of knowledge management. *Advances in Developing Human Resources* 2, 38–62.

Chapter 3

The Lessons
Learned Process

Key Learnings:

- The need to evaluate projects
- Adult Learning Theory
- Individual, Group, and Organizational Learning
- Determining Needs and Wants
- The LL Ten-Step Process

The Need to Evaluate Projects

Let's face it, most people know there is a need to evaluate projects. However, at the end of the day what matters is *results* and what we actually have time to *accomplish*. Most of us have competing demands for our time, money, and resources, which creates a dilemma of what to do with a project postmortem. In one sense, there may be a reluctance to conduct an after-action review at all. While, at the same time, there may be a temptation minimize the amount of effort.

It may be tempting to take the Path of Least Anticipated Resistance (POLAR) with respect to intraproject learning. POLAR is a bad path to follow for many reasons. If Lessons Learned (LL) are not conducted on a project, valuable knowledge may be lost forever.

POLAR icicles represent triggers that could encourage or result in shortcuts or cheats:

- **Shortcuts:** Abbreviated activities that are shortened due to constraints, resulting in early drawn conclusions; making assumptions not validated by actual lessons.
- **Cheats:** Use of findings that are not based on actual lessons to make a point.

Lessons and the Adult Learning Theory

Learning is the process by which people acquire knowledge and understanding (Marqquardt, 1996). Hopefully, this results in improved behavior or actions, resulting in enhanced processes and increased productivity. A lesson is a period of time where learning is intended to occur. The term *lesson* can have different meanings depending on the context. For example,

- **Hard lesson:** Impactful experiences that potentially result in change of behavior.
- **Key learnings:** Primary thoughts that are transferred from a review of content.
- **Lesson plan:** A detailed outline and description of an educational activity.
- **Parable:** A short narrative that illustrates a moral or ethical principle, i.e., New Testament.
- **Scenario:** A plausible and often simplified example of how a future situation.
- **Story:** A message that tells the particulars of a course of events.
- **Takeaways:** Content that provides a summary of information covered.

The Adult Learning Theory (ALT) was pioneered by Malcom Knowles. Since its inception, many practitioners have conducted supporting Research, Measurement, and Evaluation (RME) to validate its legitimacy. ALT is a philosophy that adults learn most effectively when under certain conditions. Speck (1996) highlighted some considerations for ALT, which have direct application to LL:

- Adults need to participate in small-group activities during the learning.
- Adult learners come to learning with previous experiences and knowledge.

- Transfer of learning for adults is not automatic and requires facilitation.
- Adult learners need direct, concrete experiences in which they apply learning to their job.
- Adults need feedback on how they are doing and the results of their efforts.

Individual, Group, and Organizational Learning

Individual learning is limited to engagements of personal development that can take the form of:

- **Self-study:** A form of autodidacticism that involves basically read and understand education without the interaction of another person. It may involve self-directed learning in which the learner chooses his/her own path, determines goals, objectives, etc.
- **Self-awareness:** Sensory engagement (i.e., hearing, tasting, seeing, smelling, or feeling) for the purpose of experiencing. It may not require any goals or objectives, but rather exposure to support understanding.

Group learning is designated for instances of professional development and can take the form of:

- **Instructor-led training:** Teacher-to-group instruction designed to disseminate information in an interactive fashion for the purpose of increased knowledge transfer.
- **Skills training:** On-the-job training where the trainee demonstrates knowledge provided by the instructor. They ask questions and verify concepts covered during training.
- **Competency-based training:** A form of qualification training that is situation-specific.
- Organizational learning (OL) should address personal and professional development. It includes:
- Process by which an organization acquires the skills, knowledge, and abilities to support employee growth, continuous improvement, and adapt to change.
- Culture that endorses a mindset of future-thinking, survival-oriented, competitive, market-driven, and responsive.
- Foundation by which LL produces optimal results due to the socio-political infrastructure, sponsorship, and supporting stakeholder alignment.

Journaling to Capture Learning

Journaling can be used to capture all types of learning whether it's individual, group, or organizational. Journals can be handwritten if necessary, but eventually should be converted to an electronic format. According to Loo (2002), reflective learning journals have usefulness in promoting both individual and team performance in the context of Project Management (PM) training and team building. It is interesting to note that in Loo's research, 22 percent of respondents said they found using a learning journal very difficult. As a solution, it is recommended that managers serve as coaches and help Project Team Members (PTMs) understand how to use learning journals to capture LL.

Determining Needs and Wants

Central to LL are NEEDS:

- **N**ecessity: Rationale for PE
- **E**xperience: History, expertise, environment, and of conducting PE
- **E**conomics: Cost, contracts, resources, and time required to perform PE
- **D**ocumentation: Supporting success stories realized as result of utilizing PE
- **S**ystems: Repositories and computers to support PE

Conversely, WANTS to LL could be described as:

- **W**ish lists: Futures that can be incorporated in new or subprojects for PE
- **A**dd-ons: Items that may cause scope creep due to resource requirements for PE
- **N**ice to haves: Items that will not cause scope creep to PE
- **T**rinkets: Bells and whistles that enhance the appearance of PE
- **S**pecialties: Unique factors that highlight or distinguish PE

Lessons Learned Process

At its most simplistic level, the LL process involves:

- **Noticing:** Seeing lessons that need to be captured
- **Gathering:** Capturing the lessons
- **Storing:** Saving the lessons captured
- **Sharing:** Distributing lessons

LESSONS LEARNED PROCESS

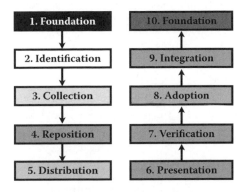

Figure 3.1   Lessons Learned process flow.

However, the comprehensive process for LL involves ten stages as illustrated in Figure 3.1.

The rationale for having ten steps instead of four is based on the following observations:

- Gaps are closed through a comprehensive process.
- Distinct steps support procedural management of lessons.
- Lessons are being handled formally as a project or subproject.
- Lessons that involve more complexity can be supported.
- Assignment of tasks can be performed at the step level.

Foundation

The foundation step is a resourcing function, which involves locating and acquiring resources. Locating resources involves observation and networking to find the right:

- **People:** PTMs, subject matter experts, and stakeholders
- **Systems:** Hardware and software for LL, i.e., repositories
- **Facilities:** Buildings and rooms
- **Equipment:** Tools, machines, and transportation
- **Materials:** Supplies and electronics devices

Acquiring resources involves reviewing time availability and cost, and may involve:

- Shared use of internal resource by part-time or full-time employees
- Use of external resources, i.e., consultants as facilitators
- Cost analysis to show return on investment (ROI) or other realized value

Issues:

- How should PTMs be selected?
- How should virtual team members be assigned?
- How should funding for the project activity be defined?

Considerations:

- Based upon availability and level of commitment.
- Consider them part of the project or subproject.
- Funding for LL can be done as part of the project.

Identification

The identification step is essentially a hunting activity to locate potential lessons, which involves observing and selecting. Observation is the act of finding lessons using all senses:

Seeing
Feeling
Hearing
Smelling
Touching
Tasting

Selection is the act of choosing lessons from those that are found and may be based on:

- Perceived impact to the organization, i.e., compliance concerns or sense of urgency
- Political drivers of what is important to customers or stakeholders

Issues:

- There is a vast amount of information that occurred on a project.
- Some of what has occurred may not be lessons, but rather extraneous information.
- Project complaints may surface as lessons and come in the form of emails.

Considerations:

- Outline applicable categories of lessons.
- Create a separate sheet for items that do not appear to be lessons.
- Complaints may be valid and worth collecting as lessons.

Collection

The collection stage is a harvesting activity to obtain lessons, which involves classifying and gathering. Classifying is the strategic act of categorizing and gathering is the tactical activity.

Think of it as a traffic signal:

- **Green:** Done right (as planned)
- **Yellow:** Done differently (another plan)
- **Red:** Wrong (not as planned)

Kendrick (2004) recommends collecting the positive aspects of the LL first. This includes passive and active collection. Passive collection occurs as unstructured processes where lessons and nonlessons are retrieved through various means, i.e., comment boxes or blogs. Active collection occurs as a structured process where lessons meeting prescribed criteria are collected.

Issues:

- Capturing lessons is difficult in many situations.
- Some people are reluctant to share lessons.
- PTMs are not co-located.

Considerations:

- Think about using a capture form for a workshop to make process consistent.
- Let people know their comments are anonymous by using a Web-based survey.
- Optimize the use of technology to provide for sharing of lessons.

Reposition

The reposition stage is a warehousing activity to save collected lessons, which involves organizing and storing. Organizing includes sorting lessons and distinguishing lessons from nonlessons. For example, project "ABC" involves the implementation of a computer network, and a not applicable lesson "456" was submitted by Jane Doe regarding her experience with laboratory equipment in a manufacturing facility. Lessons "456" may have applicability for another LL project and should be sorted accordingly. *Note*: A Lessons Learned Repository Recycle Bin (LLRRB) should be created in a shared folder for these items.

Storing lessons requires defining file names. Standards and file naming conventions are provided by many sources on the Internet, such as the Digital Projects Advisory Group. Electronic (digital) storage of lessons involves saving of lessons to an electronic repository (i.e., MS SharePoint® file server). Paper-based storage of lessons may require the use of file cabinets. Whenever practical, lessons should be maintained in a computer.

Issues:

■ There must be a relationship between PTMs and Information Systems (IS) department.
■ The size and amount of information is important to make LL useable.

Considerations:

■ Create a Good Business Practice (GBP) for LL based on best practices.
■ Determine a Standard Operating Environment (SOE) for LL.

Distribution

The distribution stage concerns sharing deposited lessons, which involves disseminating and delivering. Dissemination involves circulating lessons to identified stakeholders. A stakeholder relationship grid can track the appropriate level of stakeholder interest. The matrix should be comprehensive and developed for each Project Group (PG) and Knowledge Area (KA).

To support the instantaneous delivery of lessons, a computerized on-demand system is essential. Stakeholders should understand the purpose of the lesson. The delivery system could support notification through e-mail, and can utilize a push and/or pull system:

■ **Push system:** Notifications are sent out informing users documents are available.
  – The lessons may be distributed via e-mail.
■ **Pull system:** No notifications are sent out to users.
  – Users access lessons from a repository.

Issues:

- Are stakeholders interested in some or all lessons?
- What is the interest level in specific lessons?
- When should information be provided?

Considerations:

- Stakeholders need to be classified based on topic area.
- Create a questionnaire to determine interest area.
- A schedule for lesson distribution should be created based upon demand.

Presentation

The presentation stage involves a public relations campaign for lessons, which requires review and discussion. The purpose of the review is to address the project's vision, goals, and objectives. Discussion is the content that requires facilitation, and sometimes motivation, which may create debate, but hopefully not arguments. As a facilitator, the PTM serves as an intermediary. Motivation involves brainstorming activities and discussion should clarify and illuminate lessons.

To support coordination and discussion, a quality presentation must be prepared. Consideration should be given to every aspect of the presentation. For example, Mucciolo (2003) points out the significance of selecting colors to use within a presentation. Kodukula and Meyer-Miller (2003) emphasize public speaking skills. Mills (2007) outlines how to use PowerPoint®.

A few issues include:

- I get nervous speaking in front of my co-workers about sensitive issues.
- I have problems getting people to attend LL sessions.
- When people get to the LL sessions, they are negative and don't want to talk.

Here are some considerations:

- Express to them their involvement is valued.
- Obtain sponsorship from senior management to make attendance mandatory.
- Serve refreshments and begin on a positive note.

Verification

The verification stage is a rectifying activity, which involves revising and confirming lessons that have been presented. Verification is confirmation through examination and demonstration of objective evidence that specified requirements have been met. It is different than validation, which purposes to establish qualification gates, i.e., installation (IQ), vendor (VQ), operational (OQ), and performance (PQ). In the schema, verification typically comes after validation on a cycle, i.e., every two years.

Revising lessons is a progressively elaborative process. For example, a contributor to lesson may provide supporting details, such as video interview as supporting knowledge. Therefore, revisions to lessons become an integral part to ensure the accuracy and completeness. Confirmation serves as a checkpoint that tracking has occurred for revised lessons. It is traceability, a sign off, or versioning of lessons to support change control and management.

Issues:

■ I am not very good with statistics.
■ Who can help with analysis if I am not familiar with content?
■ What tools should I use for analysis?

Considerations:

■ Most computations performed will most likely be simple math.
■ Subject matter experts.
■ Checklist.

Adoption

The adoption stage involves agreeing and accepting verified lessons. Agreeing means stakeholders come to a mutual understanding that the lessons are accurate as presented. Acceptance means lessons are received willingly. It is an admission of something being done right, wrong, or could have been done differently. Acceptance is also affirmation that the lesson can be utilized.

Lessons that are not adopted are fostered. Metaphorically speaking, fostered lessons are taken in and cultivated or nurtured, but have not yet become part of the family of lessons. There are a number of reasons for fostering lessons:

■ It is unclear if it is a lesson or nonlesson.
■ Required supporting data or information has not been acquired to support lesson.

- Value of the lesson has not been determined and requires more review.

Issues:

- Naming convention for the files can't be determined.
- Descriptive directory structures for folders are confusing.
- Version control of lessons has not been established.

Considerations:

- Make it descriptive and include date, i.e., initiating_1_john_doe_01-01-2011.
- It could be set up by PG and KA.
- Version-specific using brackets {}, [], or () can help delineate lesson.

Integration

Integration purposes to combine, synergize, harmonize, or standardize lessons into an identified process, procedure, policy, or practice with the intent of making it better. The integration stage involves publicizing and utilizing lessons that have been adopted. Publication of lessons may involve a declaration in a company's newsletter or an announcement in an association's Web site. It must not fall short of its purpose to spotlight and reinforce best practices.

The concept of Evaluation Use dictates the purpose of evaluation is utilization of the findings, rather than to make a determination of merit (quality), worth (value) or significance (importance) for the sake of evaluation (i.e., purposes of curiosity to gain insight). Following the path of Evaluation Use, using lessons has two purposes: summative and formative. In some instances, the same lesson can be used to support both summative and formative requirements. Using lessons in a summative sense can measure impact, results, or accountability. For example, case studies become a natural output of summative evaluation. Conversely, using lessons in a formative manner can support process improvement, strategic planning, or system enhancements. For example, an organization may desire to revise policies, procedures, or work instructions.

Issues:

- I can't decide on a naming convention for the files.
- I can't determine a good directory structure for folders.
- I can't determine how to version control LL.

Considerations:

- Make it descriptive and include date, i.e., initiating_1_john_doe_ 01-01-2011
- Make it by process group and knowledge area component.
- Make it version-specific by putting version number in brackets {}, [], or ().

Administration

The administration stage involves maintaining and managing lessons that have been integrated.

Maintaining lessons involves the technical support to ensure proper functioning of the Lessons Learned Repository (LLR):

- Conformance to requirements for file types and sizes
- Adherence to records retention schedule for archiving
- Version control of submissions

Managing lessons is an oversight function to ensure the LLR is meeting the goals and objectives of the organization. The administrator should support PTMs to ensure the LLR is being used as intended. This can be accomplished not only by looking at system access statistics, i.e., file uploads and downloads, but also by speaking with LLR subscribers regarding how it is working for them. Seningen (2004) recommends LL meetings on a weekly basis, and suggests using audio or video in addition to meeting minutes to capture more detailed information.

A few issues include:

- We don't have anyone who has volunteered to manage LL.
- There is a challenge getting buy-in for specific lessons.
- What vehicles are good to gain visibility for lessons.

Here are a few considerations:

- This may be a good development opportunity for an employee.
- It is a management challenge that needs to be correctly positioned.
- Brown bag lunches.

References

Kendrick, T. 2004.
Kendrick, T. 2004. The Project Management Tool Kit: 100 Tips and Techniques for Getting the Job Done Right. Saranac Lake, NY: Amacom.

Kodukula, P., and S. Meyer-Miller. 2003. *Speak with power, passion and pizzazz.* Tucson, AZ: Hats Off Books.

Loo, R. 2002. Journaling: A learning tool for project management training and team building. *Project Management Journal.* Philadelphia, PA: Project Management Institute.

Marqquardt, M. 1996. *Building the learning organization: A systems approach to quantum improvement and global success.* New York: McGraw Hill.

Mills, H. 2007. Power Points!: How to Design and Deliver Presentations That Sizzle and Sell. New York, NY: Amacom.

Mucciolo, T. & Mucciolo, R. (2003). Purpose, Movement, Color: A Strategy for Effective Presentations. New York, NY: MediaNet.

Seningen, S. 2004. Learn the value of lessons learned. The Project Perfect White Paper Collection. Retrieved on August 11, 2011 from www.projectperfect.com.au/downloads/.../info_lessons_learned.pdf

Speck, M. 1996. Best practice in professional development for sustained educational change. *ERS Spectrum* (Spring), 33–41.

Lessons That Apply to This Chapter

1. The PT should address needs for lessons at the beginning of a project.
2. Finding lessons should involve transparent communication within all parts of the organization.
3. A variety of methods should be employed to obtain lessons. Wherever feasible, electronic methods should be employed.
4. File names and conventions must be established for the repository and published so that contributors understand the logic.
5. Some people prefer a push system where lessons-related communication is sent to them. Other people will prefer a pull system, which allows them to access lessons at their convenience.
6. Presentation skills are important to effectively convey messages.
7. If a lesson is worth collecting, it is worth verifying.
8. People are usually ready and willing to accept lessons that have value.
9. Publicizing lessons can be done in a variety of ways, from text messages to billboards.
10. Administration of lessons requires a dedicated person with attention to detail.

Suggested Reading

Abramovici, A. 1999. Gathering and using lessons learned. *PM Network*, October, 61–63. Philadelphia, PA: Project Management Institute.

Bucero, A. 2005. Project know-how. *PM Network*, May, 22–23. Philadelphia, PA: Project Management Institute.

Digital Projects Advisory Group. 2008. Guidelines on file naming conventions for digital collections. Online at: http://ucblibraries.colorado.edu/ systems/digitalinitiatives/ docs/filenameguidelines.pdf. (accessed January 14, 2011).

Knowles, M. 1998. *The adult learner: The definitive classic in adult education and human resource development.* Burlington, MA: Gulf Professional Publishing.

Ladika, S. 2008. By focusing on lessons learned, project managers can avoid repeating the same old mistakes. *PM Network*, February.

Leake, D., T. Bauer, A. Maguitman, and D. Wilson. 2000. Capture, storage and reuse of lessons about information resources: Supporting task-based information search. Online at: http://66.102.1.104/scholar?hl=en&lr=&q= cache:iyNCO80Y-OsJ:www. cs.indiana.edu/l/www/ftp/leake/leake/p-00-02.pdf+ (accessed July 15, 2007).

MacMaster, G. 2000. Can we learn from project histories? *PM Network*, July. Philadelphia, PA: Project Management Institute.

Newell, S., M. Bresnen, L. Edelman, H. Scarbrough, and J. Swan. 2006. Sharing knowledge across projects: Limits to ICT-LED project review practices. *Management Learning* 37, 167.

Pitagorsky, G. 2000. Lessons learned through process thinking and review. *PM Network*, March, 35–38. Philadelphia, PA: Project Management Institute.

Reich, B. 2007. Managing knowledge and learning in IT projects: A conceptual framework and guidelines for practice. *Project Management Journal* (June), 5–17.

Rowe, S., and S. Sikes. 2006. *Lessons learned: Taking it to the next level.* Paper presented at the PMI Global Congress Proceedings, Seattle, WA, (month and days) Project Management Institute.

Snider, K., F. Barrett, and R. Tenkasi. 2002. Considerations in acquisition lessons-learned system design. *Acquisition Review Quarterly* (Winter), 67–84.

Spilsbury, M., C. Perch, S. Norgbey, G. Rauniyar, and C. Battaglino. 2007. Lessons learned from evaluation: A platform for sharing knowledge. Online at: www.unep. org/eou/Pdfs/Lessons%20Learned%20rpt.pdf (accessed March 31, 2007).

Stephens, C., J. Kasher, A. Walsh, and J. Plaskoff. 1999. *How to transfer innovations, solutions, and lessons learned across product teams: Implementation of a knowledge management system.* Philadelphia, PA: Project Management Institute.

Stewart, W. 2001. Balanced scorecard for projects. *Project Management Journal* (March), 38–53.

Terrell, M. 1999. Implementing a lessons learned process that works. Paper presented at the 30th Annual Project Management Institute Seminars & Symposium. Philadelphia, PA: Project Management Institute.

Weber, R., D. Aha, H. Munoz-Avila, and L. Breslow. 2000. An intelligent lessons learned process. Paper presented at the Twelfth International Symposium on Methodologies for Intelligent Systems (ISMIS 2000), 358–367.

Wheatley, M. 2003. In the know. *PM Network* (May), 33–36.

Whitten, N. 2007. In hindsight: Post project reviews can help companies see what went wrong and right. *PM Network*, 21.

Williams, T. 2007. *Post-project reviews to gain effective lessons learned.* Philadelphia, PA: Project Management Institute.

Williams, T., C. Eden, F. Ackermann, and S. Howick. 2001. The use of project postmortems. Paper presented at the Project Management Institute Annual Seminars and Symposium. November. Philadelphia, PA: Project Management Institute.

Chapter 4

Lessons Learned Support Systems and Repositories

Discussion Topics:

- Lessons Learned Support System (LLSS)
- Lessons Learned Repository
- Records Management

What Is a Lessons Learned Support System?

An LLSS is inherently a system of records that are linked to a management function, i.e., Knowledge Management (KM) that is available to provide Project Evaluation (PE)-related content for a Lessons Learned Report (LLR). These systems must be accessible to Project Team Members (PTMs) from a reporting perspective so the information is transferable to the LLR. While most systems are computerized, paper-based (manual systems) should always be considered. Many LLSS are adaptable to meet the specific LL needs of the organization, i.e., through a reporting tool, such as business objects. An LLSS can range in application, size, and complexity. The key is to be able to locate the LLSS and then to determine the most appropriate use for LL (Table 4.1).

Table 4.1   Lessons Learned Support System by Management Category and Function

Management Category	Function	Examples
Records	Organizes file space and may incorporate workflow support	Standard Operating Procedures (SOPs) to ensure compliance
Project	Manages cost, resources, and time across the project life cycle	Project, program, and portfolio management reports
Human Resource	Tracks staff productivity	Job evaluations
Supply Chain	Supports logistics for products and services delivery and return	Inventory kept on hand for freshness dating
Sales	Sells products or services	Number of sales this month
Marketing	Advertises products or services	Promotion effectiveness
Public Relations	Voice company message	Response to a news report
Knowledge	Encourages organizational learning and growth	Decision support using business intelligence shown in dashboards
Customer Relationship	Tracks sales and marketing interactions	Postsales customer history with technical support required
Quality	Ensures adherence to standards	Corrective/preventative action
Operations	Oversees the overall operations	Address staff interactions
Finance	Records expenditures, revenues and investments	Return on investment
Procurement	Negotiates vendor contracts	Improved risk management
Information Technology	Supports computer-based connectivity and communication	Software utilization

Table 4.1 *(Continued)*   **Lessons Learned Support System by Management Category and Function**

Management Category	Function	Examples
Research and Development	Engages discovery, trial and delivery	Laboratory and manufacturing
Facilities	Maintains building and supporting infrastructure	Space management
Engineering	Designs and creates systems	Cost to produce new building
Environmental, Health and Safety	Monitors work practices	Reduction in injuries due to improved office ergonomics
Food Service	Supplies meals, snacks, and beverages	Number of meals served on site and cost per employee
Security	Enables employee access and respond to immediacy	Response time to fire-related emergencies
Shipping/ Receiving	Delivers mail and packages	Cost of packages sent out
Graphics and Copying	Creates images and makes copies	Number of projects completed
Medical and Wellness	Improves physical health of staff	Number of sick days taken
Legal	Addresses regulatory or statutory issues	The number of complaints resolved during a given period
Learning Management	Trains and develops people within the organization	Kirkpatrick Level 3– Behavior Evaluation; extent of applied learning on the job (transfer)

LLSS, in some cases, may be departments, i.e., Human Resources Information System (HRIS) and, in other instances, will be initiatives, i.e., Customer Relationship Management (CRM). Whether it is a department or an initiative, both are classified under management in Table 4.1. The key to these functional areas of management is whether a supporting system, i.e., to measure customer satisfaction, is integrally involved in KM. Where a system is involved, content can be potentially extracted to support LL. Because LL may be handled as a subproject, it is important to establish parameters for the content that will be retrieved when using a variety of systems.

Why Use a Lessons Learned Support System?

An LLSS contains source files and, therefore, a preferred method to obtain information. For example, a company may desire to move to a paperless environment. They embark on a journey to determine how to reduce the cost associated with printing of documents on copiers. So, they undertake a specific project that involves LL. While there may not be a request for new copiers, LL can have instrumental value in supporting a Utilization–Case–Study (UCS). In a UCS, the goal is to determine the amount of use a particular item has on a target. Considerations for copier use may include:

- Number of pages printed
- Amount of toner consumed
- Cost of electricity
- Cost of paper ordered
- Percentage of paper printed that is retained versus discarded
- Mean time to repair copier and mean time between failure
- Number of employees using copier

The above factors are quantitative. However, as a part of the LL process, qualitative factors also should be considered, such as:

- Types of documents being printed, i.e., compliance documents requiring storage
- Transportability, i.e., people needing hard copies for offline review
- Quality, i.e., clarity of printouts versus size of pixels on screen
- Mark-up requirements, i.e., ability to comment on paper versus view only on screen
- Retention, i.e., need to keep a printed copy on hand for recordkeeping purposes

What Is a Lessons Learned Repository?

An LLR is a system that is designated to support LL in terms of:

- Collection: Allowing for submission of lessons
- Storage: Saving of files
- Retrieval: Accessing of files
- Distribution: Sending of files
- Administration: Maintenance of files
- Presentation: Conversion of files to another format to support visualization

An LLR may be configured in a variety of ways:

- Shared folder system: To support organization of content
- Database: Structured with defined fields to support queries
- Formatted Web pages: Containing a series of defined fields
- Unformatted Web pages: Containing minimal fields, i.e., blog
- File cabinet: Paper-based documents arranged sequentially or by topic area

How are system requirements determined? LLR requirements are uncovered through a Systems Needs Assessment (SNA). The SNA is a very important task. To ensure the success of LL system implementation, address LL system requirements formally. Start with a project charter, then develop a solid scope statement. A well organized project plan should be constructed, outlining the Project Life Cycle (PLC), initiation through closure. The Knowledge Area (KA) activities, communications through time, also should be included in the project plan. The sponsor and key stakeholders should be involved early in the process.

Computer-Based Repositories versus Paper-Based Systems

While there may be regulatory requirements to maintain paper records to support audit trails, for the purpose of LL, a computerized LLR is preferred over a paper-based repository for many reasons. To support the rationale of moving from paper to computer, the Project Team (PT) should thoroughly investigate the features, functions, and benefits of an electronic LLR.

Digital versus Electronic

The terms digital and electronic are frequently used interchangeably and in some cases are synonymous. In other cases, digital and electronic are somewhat different. For example, a digital signature may be a scanned signature of a person's

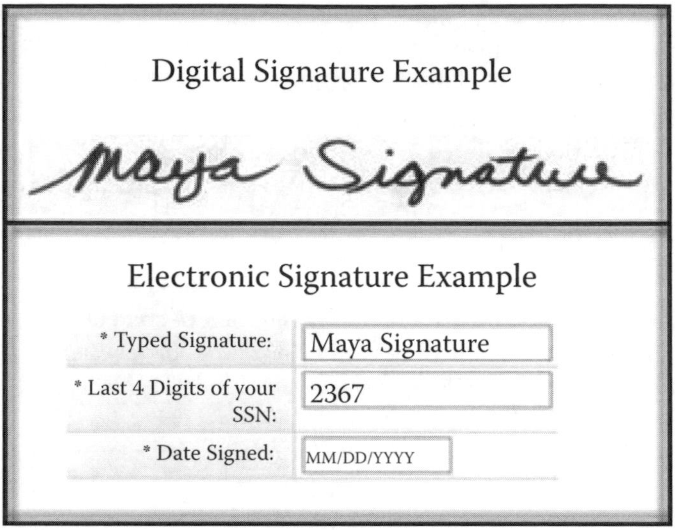

Figure 4.1   Digital versus Electronic Signature

handwriting that is placed within a memo or letter (Figure 4.1), whereas, an electronic signature may be a typed in name "Josephine Z. Public" within a form field entry, which is accompanied by a checkbox stating "I agree," then followed by clicking an Accept button. The trend is to use digital signatures for a personalized representation, i.e., for acknowledgement purposes and electronic signatures for compliance purposes to ensure traceability. A company president will use their scanned signature to thank employees in a letter of recognition. On the other hand, a healthcare company will use electronic signatures to route approvals on a requisition.

Features of a Computer-Based Repository

First, a computer-based storage technology offers significant benefits over paper-based file cabinets:

- Accessibility: send and receive content 24/7
- Capability: Functions an electronic system can perform in terms of report generation
- Dependability: System can be backed up or mirrored to support business continuity
- Expandability: Increase of file space can be performed
- Feasibility: Likelihood that the system will be used as intended

- Identity: Repository becomes a recognized system of record
- Likeability: Repository must be accepted by the community that will use it
- Measurability: Metrics are available regarding system usage at all levels
- Productivity: Work that be accomplished through the system
- Quality: User-friendly interface that meets established criteria
- Security: Able to set permissions regarding level of access
- Transportability: System relocation can occur as needed
- Usability: Simultaneous real-time support for subscribers
- Verifiability: Tracking what has occurred through system use

Functions of an Electronic Repository

Second, a computerized LL system should ideally have the following functions:

- Down-loadable: Retrieve lessons
- Examinable: Support audits or evaluations
- Flexible: Supports multimedia, i.e., e-mails, graphics, audio, and video
- Manageable: Allow for timely administration
- Queryable: Allowing for search based on keyword
- Serviceable: Allowing for routine maintenance
- Trainable: Offer online help, technical support, or related job aid
- Uploadable: Contribute content
- Viewable: Allowing content to be displayed

Benefits of an Electronic Repository

Third, the organization should realize the following as the result of implementing an LLR:

- Allocated: System use, space requirements, administrative support, etc. can be determined
- Collaborated: New ideas and strategies can be initiated based on lessons contributed
- Documented: System exists that supports transfer of knowledge through LL
- Established: Sponsored, visible, and recognized
- Formatted: Set up in a way to support standardization of content
- Integrated: Becomes a part of the organization's knowledge assets
- Leveraged: Used to support project life-cycle phases
- Networked: Can support internal and/or external contributions
- Optimized: Designed specifically to support the interchange of lesson

Records Management

A record is stored data that contains completed, current, or future information about an organization's activities, i.e., transactions. Records are documentation that may be in paper and/or an electronic format. A paper record becomes electronic when it is transmitted into a computer, i.e., through scanning. A record becomes electronic when it is saved in a digital format, i.e., Adobe Acrobat® PDF. An e-signature is a legally binding, digital representation of a person's handwritten signature that becomes part of the record it acknowledges.

Records Management (RM) involves identification, classification, preservation, destruction, and control of an organization's records regardless of the storage medium. ISO 15489: 2001 standard defines RM as "the field of management responsible for efficient and systematic control of the creation, receipt, maintenance, use and disposition of records, including processes for capturing and maintaining evidence of and information about business activities and transactions in the form of records."

LL are a part of an organization's records and may be subject to records management policies. Therefore, it is important to consider the Records Life Cycle (RLC) for LL (Figure 4.2). When a lesson is captured, there may be a tendency to hold on to it forever because of its perceived instrumental value. However, that may not be practical. There are circumstances when a lesson becomes no longer useful, i.e., advancements in technology that cause previous ways of doing something to become obsolete. Therefore, all lessons should be maintained with the appropriate controls to ensure they apply.

This life cycle involves four stages:

1. Capture
2. Access
3. Retain
4. Delete

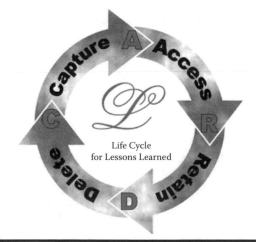

Life Cycle
for Lessons Learned

Figure 4.2

Digital or electronic encompasses to the overarching technology and methodology used to store lessons on a computer in the form of:

- **Raw Data:** unprocessed files or unstructured information that can be interpreted by a specific program or utility
- **Files:** records meeting specific criteria that are saved in a user accessible format, i.e., documents, spreadsheets, databases, audio, video, etc.
- **Images:** pictures, graphics diagrams, charts, illustrations, etc. that have been saved in a user accessible format.

Knowledge Management

Knowledge Management (KM) is the practice of locating, capturing, synthesizing, storing, and sharing observations and experiences to support the development of intellectual capital. The goal of KM is continuous improvement of an organization's processes, i.e., their product and services. KM is an overarching concept that extends beyond RM and includes areas such as:

- **Enterprise Decision Management** involves automated decision design and support (i.e., business intelligence) to manage interactions with customers, employees, and suppliers.
- **Information Management** is the high-level, strategic process of enabling people to utilize systems in support of content retrieval, storage, and distribution. Information Life Cycle Management (ILM) is an Information Technology (IT) strategy. Phase 1 involves establishing a tiered architecture for ILM; phase 2 involves bringing ILM and tiered storage to a key application; and phase 3 involves extending ILM across multiple applications.
- **Idea Management** is an approach to managing innovation by putting systems in place to organize ideas. It involves the evaluation of concepts and review of LL in a structured fashion to support the selection of the best ideas with the highest potential for implementation. It is a five-stage process that supports product development: (1) Generation, (2) Capture, (3) Analysis, (4) Implementation, and (5) Review.
- **Value Stream Management (VSM)** is a Supply Chain Management (SCM) methodology that evaluates the flow of goods and services from the supplier to the customer. It embodies LL by continuously reviewing customer requirements, inventory, warehousing, procurement, distribution, and transportation through Enterprise Resource Planning (ERP). The goal of VSM is to support just-in-time lean initiatives to optimize the SCM network.

Lessons That Apply to This Chapter

1. Organizations need to define their knowledge assets and determine which ones will provide support for LL.
2. Conducting LL on an LLSS may be an initiative in itself that becomes a project.
3. The PT must develop relationships with system owners throughout the organization so that they become a trusted resource for knowledge management.
4. Records management is a serious concern and care must be taken regarding temporary records before they become permanent.
5. Permanent records are subject to legal review and may contain sensitive information that must be controlled.
6. A records retention schedule must be put in place for an organization.
7. An LLR must adhere to records retention practices and schedules, which can be done in a number of ways through archival or deletion.
8. Lessons that contain proprietary, confidential, or sensitive information should be restricted and password protected.
9. Business continuity plans should be established for an LLR.
10. Metadata is an overarching concept that needs to be reviewed continuously. The LLR administrator needs to know who is creating what, when it is created, and where it is stored.

Suggested Reading

Bates, S., and T. Smith. 2007. *SharePoint 2007 user's guide: Learning Microsoft's collaboration and productivity platform.* New York: Springer-Verlag.

Borghoff, U., and R. Pareschi. 1998. *Information technology for knowledge management.* Heidelberg, Germany: Springer-Verlag.

Cohen, D., L. Leviton, N. Isaacson, A. Tallia, and B. Crabtree. 2006. Online diaries for qualitative evaluation: Gaining real-time insights. *American Journal of Evaluation* 27, 163–184.

Cronholm, S., and G. Goldkuhl. 2003. Strategies for information systems evaluation—six generic types. *Journal of Information Systems Evaluation* (electronic) 6 (2): 65–74.

Hall, H. 2001. Input-friendliness: Motivating knowledge sharing across intranets. *Journal of Information Science* 27, 139–146.

Kaiser, S., G. Mueller-Seitz, M. Lopes, and M. Cunha. 2007. Weblog-technology as a trigger to elicit passion for knowledge. *Organization* 12, 391–412.

Kaner, C., J. Bach, and B. Pettichord. 2002. *Lessons learned in software testing.* New York: John Wiley & Sons.

Rist, R., and N. Stame. 2006. *From studies to streams: Managing evaluative systems.* New Brunswick, NJ: Transaction Publishers.

Sikes, D. 2000. Using project websites to streamline communications. *PM Network*, June, 73–75.

Tiwana, A. 2002. *The knowledge management toolkit: Orchestrating IT strategy and knowledge platforms*. Upper Saddle Ridge, NJ: Pearson Publications.

Walsham, G. 2002. What can knowledge management systems deliver. *Management Communication Quarterly* 16, 267–273.

Weber, R., D. Aha, and I. Becera-Fernandez. 2001. Intelligent lessons learned systems. *International Journal of Expert Systems Research & Applications* 20 (1): 17–34.

Webster, B., C. Hare, and J. McLeod. 1999. Records management practices in small and medium-sized enterprises: A study in NE England. *Journal of Information Science* 25, 283–294.

Yusof, Z., and R. Chell. 2000. The records life cycle: An inadequate concept for technology-generated records. *Information Development* 16, 135–141.

Chapter 5

Best Practices and Benchmarking

Discussion Topics:

- Communities—of Interest, Practice, and Development
- Best Practices
- Benchmarking

Communities of Interest

A Community of Interest (COI) is a virtual meeting place (Web site) where people share a common interest, e.g., a hobby. Members randomly share lessons pertaining to general topics. Participation in a COI is usually free (or at a low cost), informal, and can be entertaining. It purposes to create a feeling of belonging and friendship. There are few rules or protocols that govern a COI. However, it is expected that members will use appropriate language and respect diversity of thought, culture, and walks of life. Examples of Web-based COIs are MySpace created in 2003, FaceBook introduced in 2004, and YouTube launched in 2005.

Communities of Development

A Community of Development (COD) is a virtual meeting place where participants are focused on a specific topic to support personal or professional development. They share lessons in forums in a manner referred to as threads. Participants do not have to be members and are primarily concerned with finding a solution to a problem, i.e., how to fix code in a software program. A COD is frequently hosted by a software company in hopes of addressing technical support issues by providing a way to get answers through frequently asked questions (FAQs).

Communities of Practice

A Community of Practice (COP) is a virtual or physical meeting place where members indicate a collaborative interest in a professional activity, i.e., project. The entities are groups of people, e.g., institutions, agencies, companies, businesses, or associations. The group evolves over time because of the members' shared involvement in a particular domain or area, or it can be created specifically with the goal of gaining knowledge related to their field. Members share best practices and Lessons Learned (LL). Examples of COPs include professional associations, e.g., the Project Management Institute (PMI) and American Evaluation Association (AEA). COPS are sometimes exclusive or restricted to the people within a profession and may require licensure or membership. It is not uncommon for members to pay a fee to subscribe to COPs.

What Are Best Practices?

The American Society for Quality (ASQ) defines Best Practices (BP) as "a superior method or innovative practice that contributes to the improved performance of an organization, usually recognized as best by other peer organizations." Body (2006) defines BP as "the policy, systems, processes, and procedures that, at any given point in time, are generally regarded by peers as the practice that delivers the optimal outcome, such that they are worthy of adoption."

Kerzner (2004) defines a BP as "reusable activities or processes that continuously add value to the deliverable of the projects." Kerzner relates that acceptable BP can appear in a variety of settings, including working relationships, design of templates, and the manner in which PM methodologies are used and implemented. He also states that companies developing their own BP have greater success, particularly when they incorporate their own BP and LL from other activities.

Critics argue that many external factors impact the label of best practice, e.g., time, technology, and cost. Some of these critiques indicate that it is important to be careful of the BP label. In some institutions, BP is the appropriate label, and, in other associations, it is not. To support this argument, Patton (2001) contends that BP widespread and indiscriminate use as a term has devalued it both conceptually and pragmatically. He contends there is "pridefulness" in proclaiming that one is practicing what is best, and, therefore, one should employ the language of effective practices or evidence-based practices. Not only can a BP label be misleading to the item under study, in addition, it frequently infers a generalization that might not apply to the rest of the population under study.

Despite its widespread use and acceptance, the concept of BP should not be ignored. Damelio (1995) states that BP are those methods or techniques resulting in increased customer satisfaction when incorporated into the operation. Individuals and organizations on personal and professional levels embrace BP. Coakes and Clarke (2006) explain that a BP is determined by the stakeholders and producers and may involve many subjective criteria. For example, the speed and technique of changing tires on a racecar is not a transferable BP for a passenger vehicle. An appropriate BP label might be: "BP for changing tires on an Indy 500 race car during the Indy 500 race."

A commitment to using "good" practices in any field represents a commitment to using the knowledge and technology at their disposal to ensure compliance, i.e., Good Manufacturing Practices (GMP) or Good Laboratory Practices (GLP) as evidenced in the pharmaceutical industry. As organizations come to understand the connection between BP and LL, they should develop systems that support a transfer of knowledge.

Berke (2001) considers LL and BP as an integral process and defines it as Best Practices Lessons Learned (BPLL). Berke explains that BPLL is the building block of Organization Learning (OL). OL is a pillar in the foundation of an organization's knowledge and embodies organizational development, change, and resilience, and thereby maturity. Preskill and Catasambas (2006) define OL as the intentional use of learning processes at the individual, group, and system level to continuously transform the organization in a direction that is satisfying to its stakeholders.

Hynnek (2002) claims the majority of the effort in ensuring the success of LL in an organization is contingent upon five factors:

1. Recognize and praise good practices.
2. Maintain good practices and extend them to the entire organization.
3. Define solutions for the problems encountered by the project.
4. Generate an action plan for the implementation of these solutions.
5. Track progress on the implementation with ongoing feedback.

Capturing BPLL should answer the following questions:

1. What is the merit, worth, or significance of the lesson?
2. Is the lesson explained so that people external to the project understand it?
3. Is the context of the lesson described, i.e., a postimplementation review?
4. Will the lesson help manage a similar project in the future?
5. Does the lesson have broader application than this project?
6. Does the lesson provide new insight?
7. Does the lesson reinforce information already in place or is it contradictory?

There are six considerations to defining BP:

1. **Who**: Includes criteria, such as authorship, credibility, and experience
2. **What**: Concerns what makes it a BP and includes values, such as usefulness
3. **Where**: Looks at the situation, location, and application-specific connections
4. **When**: Outlines when the BP is engaged and involves aspects, such as implementation
5. **Why**: Addresses the uniqueness, difference, and innovativeness
6. **How**: Explores the methodology, procedure and process

What Is Benchmarking?

A **Benchmark** is a standard used for measurement purposes. The objective of benchmarking is to compare similar variables, characteristics, or attributes, i.e., performance of a product, service, or result. The goal of benchmarking is to help organizations understand how to continuously adapt their processes to remain competitive in the industry by spelling SUCCESS (Sustained Understanding Common Critical Expert Strategies). Organizations use benchmarks to develop new and improved strategies. Benchmarking as an activity can utilize LL.

Belshaw, Citrin, and Stewart (2001) say that benchmarking is an attempt to compare one organization that has demonstrated a BP against another. This definition is particularly applicable to industry benchmarking or best-in-class benchmarking depending on the context. Damelio (1995) defines benchmarking as "an improvement process used to discover and incorporate BP into your operation." He explains that what makes a **better practice** than another depends on the criteria used to evaluate the practice.

Coers et al. (2001) state that benchmarking is "the process of comparing and measuring your organization against others, anywhere in the world, to gain information on philosophies, practices, and measures that will help your organization take action to improve its performance." Mulcahy (2005) maintains, "benchmarking involves looking at past projects to determine ideas for improvement on the current project and to provide a basis to use in measurement of quality performance."

Benchmarking may be conducted internally or externally. Internal benchmarking is conducted between various and may include analyzing processes. External benchmarking occurs between organizations and may include analyzing services. Benchmarking may be performed **quantitatively, qualitatively, or mixed-method**:

- Quantitative benchmarks might focus on number of items produced.
- Qualitative benchmarks might focus on quality of items produced.
- Mixed-method benchmarks might focus on number of top-quality items produced.

Reider (2000) states benchmarking is directed toward the continuous pursuit of improvement, excellence in all activities, and the effective use of BP. There are a number of ways this can be accomplished (Bogan and English, 1994):

1. Begin with the identification of values and criteria before conducting a benchmark.
2. Next, the values and criteria are defined in a project plan.
3. A measurement approach is determined, which most closely depicts the evaluand.

An organization needs to determine the preferred approach for benchmarking, but may want to consider the following categories:

- **Internal**: Groups, teams, departments, e.g., within the same organization
- **External**: Other organizations, e.g., associations, institutions, or companies
- **Industry**: Offering a similar product or service, e.g., best-in-class
- **Role/Function**: Looking at the same tasks, e.g., job description
- **Process**: Reviewing similar steps, e.g., manufacturing procedures

Comparative versus Competitive Analysis

A comparative analysis is an item by item, or step by step, comparison of two or more comparable procedures, processes, policies, practices, or products. Its intent is to serve as a gap analysis or a deficiency assessment. For example, a gap analysis is sometimes imposed by a regulatory agency, e.g., the Food and Drug Administration (FDA). A gap analysis reviews each procedure for similarities and differences as well as potential missing items (holes), ambiguity, or errors. In the pharmaceutical industry, for example, changes in standard operating procedures (SOPs) may require a comparative analysis to support the development of revised procedures to ensure consistency in drug manufacturing. Similarly, a deficiency assessment seeks to determine met and unmet needs. For example, in healthcare, training records are very important to the Occupational Safety and Health Administration (OSHA). Each nurse maintains records of his/her training to ensure his/her education is up to date regarding medical practices. This is essential to ensuring patient safety. A syllabus must be created and maintained, which lists core training requirements for nurse recertification. A met/unmet report will identify if core training requirements have been satisfied.

A competitive analysis is a comprehensive review of two or more companies' performance characteristics (criteria-based) as it relates to their product, service, or result. It involves detection and examination of the factors that cause the difference. In a competitive analysis, Company X desires to know why Company Y has a superior product. A competitive analysis can be engaged by choice or imposed. For example, in the automobile industry, in 2010, the National Highway Traffic Safety Administration (NHTSA) proposed a new auto safety regulation that would require rearview backup cameras in all new cars sold by 2014. This change in technology, of course, will impact used car sales that do not have a backup camera installed. Car dealers may need to think competitively about installing rearview cameras in used cars to avoid loss of sales.

Choosing a Measurement Approach

An organization should choose a consistent approach to support measurement. This will drive BP and benchmarking initiatives. For example, if the procurement department is trying to analyze vendor performance, it will need to choose one of the three methods listed below:

1. **Grading:** Assigning A, B, C, D, F letter grades. Respectively, the letter "A" represents superior performance and "F" represents failing performance.
2. **Rating:** Judging something in terms of quantity or quality. Three- to five-point scales are typically used, e.g., poor, fair, good, and excellent.
3. **Ranking:** Rank ordering involves placing items in a particular order based on relative importance or achievement. Ranking can be relative, e.g., first, second, and third place.

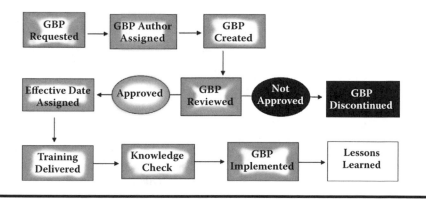

Figure 5.1   Creating Good Business Practices (GBPs).

In the above example, if the company chooses Rating for vendor evaluations, checklists and scales must be developed.

Good Business Practices and Bulletins

A Good Business Practice (GBP) is a documented guideline that an organization follows to ensure consistency in specific business operations. A GBP is a formal document that specifies policies and procedures of operational and management functions. For example, an organization may have a GBP on how it conducts internal training, i.e., self study or instructor-led.

A Good Practice Bulletin (GPB) is a periodic communication, i.e., newsletter that provides a supporting example of a GBP in the form of LL. A GPB is usually short, one to four pages. A GBP is generally a supplement to an organization's Quality Standards (QS). A QS is a framework for achieving a desired level of quality within an organization based on defined criteria. Establishing a GBP, with a supporting GPB, i.e., a quarterly edition, can significantly enhance the visibility of LL within an organization. GBP development begins by assembling a group of Subject Matter Experts (SMEs) who follow a process similar to what is illustrated in Figure 5.1.

References

Belshaw, B., S. Citrin, and D. Stewart. 2001.

Berke, M. 2001. *Best practices lessons learned (BPLL): A view from the trenches.* Proceedings from the Project Management Institute. Philadelphia, PA: Project Management Institute.

Body, A. 2006. *Principles of best practice: Construction procurement in New Zealand.* Queensland University, New Zealand: Construction Industry Council.

Bogan, C. & English, M. 1994. Benchmarking for Best Practices: Winning Through Innovative Adaptation. New York, NY: McGraw Hill.

Coakes, E., and S. Clarke. 2006. *Communities of practices and the development of best practices.* Hershey, PA: IGI Publishing.

Coers, M., C. Gardner, L. Higgins, and C. Raybourn. 2001. *Benchmarking: A guide for your journey to best-practice processes.* Houston, TX: APQC Publishing.

Damelio, R. 1995. *The basics of benchmarking.* Portland, OR: Productivity Press.

Hynnek, M. 2002. A real life approach to lessons learned. Project Management Innovations, Vol 2, pp. 5–6. Retrieved on August 11, 2011 from dewi.lunarservers. com/~ciri03/pminpdsig/.../November2002NPDSIG.pdf.

Kerzner, H. 2004. *Advanced project management: Best practices on implementation*, 2nd ed. Hoboken, NJ: John Wiley & Sons.

Mulcahy, R. 2005. *PMP exam prep: Rita's course in a book for passing the PMP exam* (5th ed.). Minneapolis, MN: RMC Publications, Inc.

Patton, M. 2001. Evaluation, knowledge management, best practices, and high quality lessons learned. *American Journal of Evaluation* 22, 329–336.

Preskill, H. & Catsambas, T. 2006. Reframing evaluation through appreciative inquiry. Thousand Oaks, CA: Sage Publications.

Reider, R. 2000. *Benchmarking strategies: A tool for profit improvement.* Hoboken, NJ: John Wiley & Sons.

Lessons That Apply to This Chapter

1. If something is worth doing, it is worth measuring.
2. COIs can be a good place to share personal lessons.
3. CODs are helpful when technical support is lacking.
4. COPs can be beneficial, especially for professional associations.
5. GBPs are essential to developing one voice in a company.
6. A GBP may help with compliance for LL.
7. Benchmarks are not as valuable if you are comparing "apples to oranges."
8. Grading can utilize +/- to account for in-between grades and then becomes known as banding.
9. Rating is easiest to translate cognitively using three- to five-point scales.
10. Ranking does not necessarily imply something is good or bad.

Suggested Reading

Adams, T. 2005. *Selecting the best method for process baselining.* Online from http://www. bpminstitute.org/articles/article/article/selecting-the-best-method-for-process-baselining.html (accessed September 29, 2007).

American Productivity and Quality Center. 2004. *Facilitated transfer of best practices—Consortium learning from best-practice report.* Houston, TX: APQC.

Grabher, G. 2004. Learning in projects, remembering in networks: Communality, sociality, and connectivity in project ecologies. *European, Urban and Regional Studies* 11, 103–123.

Harrison, W. 2004. *Best practices: Who says.* IEEE Software. Washington, D.C.: IEEE Computer Society.

Newell, S., and J. Swan. 2000. Trust and interorganizational networking. *Human Relations* 53 (10): 1287–1328.

Chapter 6

Case Studies

Key Learnings:

- This chapter summarizes a variety of case studies.
- Case studies are qualitative research involving empirical inquiry.
- Case studies are an input to and output of Lessons Learned (LL).
- Case study is sometimes used to refer to LL.
- Consider the instrumental value of the lessons and think how the transfer of lessons could be improved using the framework presented in Chapter 3.

Introduction

A case study (case) is a qualitative research method used to examine real-life situations, which can provide supporting rationale for the application of new ideas (Yin, 2008). Cases involve empirical inquiry and are used to prove theories. Cases can help Project Team Members (PTMs) understand complex issues. LL is both an input to and output of a case, whereby key lessons are elaborated upon to provide supporting details. PTMs should review case studies before conducting LL and to develop a case after completing LL. To develop a case:

- Select the issues under discussion
- Categorize issues as a topic, i.e., through process groups and knowledge areas
- Define the research questions

- Determine the lessons
- Outline data and information gathering methods
- Perform analysis
- Conduct LL

Case Studies

Bellcore

Bellcore (Bell Communications Research, a spin-off of AT&T Bell Labs) was the research arm of the Bell Operating Companies. Bellcore was purchased by SAIC in 1997. The purchase agreement required Bellcore to change its name: Bellcore became Telcordia Technologies in 1999. Bellcore was a large telecommunications conglomerate.

Bellcore desired to learn from past mistakes and to improve project management tasks correctly in the least amount of time. Sharing lessons across Bellcore was a key element of certification at level five of the Software Engineering Institutes (SEI) capability maturity model for software development. Their application of lessons learned was formative evaluation of LL. They conducted an evaluation of their LL process and determined lessons had limited use, i.e., being collected occasionally and were not being shared throughout the organization. Moreover, they found lessons focused on technical aspects of the project and did not include more business-oriented lessons. Postmortems were usually performed only at the end of the project, sometimes resulting in the loss of important details.

In response to improving LL, they engaged the Project Management Center of Excellence (PMCOE), which was a part of the Project Management Office (PMO) within Bellcore. This group began their LL research process with a comprehensive literature search to identify best practices. They determined that best-in-class methods involved observing lessons during all phases of the project life cycle, scheduling a final LL session as soon as possible after project closure and categorizing lessons appropriately were essential. In conclusion, Bellcore designed a technologically sound, online database for LL where profiles could be enhanced over time.

CALVIN

Leake et al. (2000) claim case-based reasoning systems have been developed to capture experiential knowledge to support decision making. Case-based reasoning systems support task-driven research by storing cases and recording specific information resources that users consult during their search efforts to support decision making. Users then can proactively suggest these cases as information resources to future related searches.

This new system called CALVIN (Context Accumulation for Learning Varied Information Notes) investigates how LL can support the process of finding information relevant to a project. CALVIN represents another step toward artificial intelligence-related query for use in LL systems. CALVIN learns three types of lessons.

First, CALVIN captures LL about research resources that are worth considering in a specific context. It stores cases containing information about what the resources' previous users found using similar search criteria. Second, CALVIN captures lessons that the user chooses to enter as annotations to the selected resources. Third, CALVIN is linked to a concept-mapping component for recording, clarification, and sharing of LL about important ideas in the domain, which is based on the available information in the repository.

The approach used by CALVIN is revolutionary in that it integrates problem-solving capabilities with task-relevant resources. It integrates a just-in-time, just-in-case methodology of information retrieval that mimics capabilities in artificial intelligence systems. This enables the user to spend less time on individual searches, which results in the ability to perform more searches. Adaptive search capabilities are an excellent enhancement for LL systems.

Center for Army Lessons Learned

The Center for Army Lessons Learned (CALL) was established on August 1, 1985, as a directorate of the Combined Arms Training Activity (CATA) to be located at Ft. Leavenworth, Kansas. CALL collects, stores, and analyzes data and information from a variety of sources, including army operations to produce lessons for the military. CALL is deployed worldwide to provide joint interagency, intergovernmental, and multinational (JIIM) forces with historic and emerging observations, insights, and LL.

CALL offers a five-day course on LL designed to train a target audience of officers, warrant officers, and NCOs (SSG to LTC), serving at brigade, division, corps, or equivalent levels. DAC/DoD/Other Government Entity (OGE) civilians or contractors assigned similar responsibilities may take the course provided space is available. As of FY 2011, this program is an Army Training Requirements and Resources System (ATRRS)-level course and slots must be reserved in ATRRS. This increased emphasis in LL, represents a heightened awareness in U.S. national and global security. The CALL Web site at http://usacac.army.mil/cac2/call/ll-links.asp provides links to other government LL content:

Air Force
Army
Federal
International
Joint
Marines
Navy

Eli Lilly

Eli Lilly established a shared learning program for product teams (Stephens et al., 1999) to support communication and implementation of best practices, both internal and external to optimize performance. The idea was championed by a member of senior management who assigned a project manager, knowledge manager, and organizational effectiveness person to the initiative. This focus on sharing lessons was in response to a heavy weight team initiative (one molecule per team) that involved phase 3 and 4 drug product development, requiring interfaces between project management, regulatory, manufacturing, marketing, and medical (statisticians, clinical research, data management, writers, and doctors).

Lilly understood that failure to leverage lessons in its company meant that more resources and time to complete the necessary work would be required. Lilly created interactive networks for core roles on product teams to promote innovations and lessons. Moreover, they noted that teams would likely encounter similar pitfalls and would probably reinvent the wheel. Lilly determined that not taking advantage of LL would cause the company to lose its competitive position.

The basis for Lilly's strategy was the value proposition that considered return on investment. Lilly developed solid models for the learning transfer process, which included a shared learned database to support documentation and dissemination of information to stakeholders. They invested in professional development of project managers through Project Management Institute (PMI) certification. As a result of LL, Lilly reported dramatic improvements in an enhanced cycle time for clinical trial reports and reduction in its product submission timeline for Japan. This innovation in LL enabled Lilly to introduce a leading drug in its therapeutic area ahead of the competition.

Hurricane Katrina

The 2005 Atlantic hurricane (Katrina) was the costliest natural disaster in U.S. history. In Katrina, 1,836 people died in the actual hurricane or as the result of subsequent flooding, making it the deadliest U.S. hurricane since the 1928 Okeechobee hurricane. Total property damage was estimated at $81 billion.

Because of known levee problems that remained insufficient, a lawsuit was filed against the U.S. Army Corps of Engineers, who were designers and builders of the levee system. There was also an investigation of the responses from federal, state, and local governments resulting in numerous resignations or terminations. On the flip side, the U.S. Coast Guard, National Hurricane Center, and National Weather Service were commended for their responses.

Then-President George W. Bush said, "The government will learn the lessons of Hurricane Katrina. We are going to review every action and make necessary changes so that we are better prepared for any challenge of nature, or act of evil men, that could threaten our people." While this acknowledges the potential value of LL, the current system for Homeland Security does not provide a framework to manage the challenges presented by twenty-first century catastrophic potential threats.

This is repeatedly demonstrated by the frantic responses to natural disasters, such as flooding or forest fires, rather than planning for anticipated occurrences, such as these. The goal is to prevent similar future errors. In the case of Hurricane Katrina, a collaborative LL repository should have been created to share findings that pertain specifically to disaster recovery, safety, security, health and wellness, and similar initiatives. There are countless other examples of where sharing information could result in the avoidance of disaster or substantially minimize the number of people affected. With all due respect to the hardworking people in the U.S. government and our dedicated military, there is a true need for better management of LL.

Kentucky Transportation System

An LL system is a critical component to support the design and construction of roadways and bridges. The evaluation report produced by the Kentucky Transportation System (KTS) examined the development of a centralized, Web-based Lessons Learned System (LLS) that allowed for uploading of text and attachments. The significance of LL for KTS is they found that one of the biggest changes with constructability is capturing construction knowledge to share with designers and other parties in a systematic and reliable way.

The following objectives were identified for their research:

■ Identify LLS currently used by other transportation organizations
■ Define directed LLS functional capabilities
■ Develop a system design for a LLS
■ Recommend a LLS for integration into KTC

The LL process used by KTC uncovered the following key characteristics of a LLS:

- Information gathering and capturing
- Analysis and validation
- Data entry into a knowledge base
- Best practice determination

KTC also determined that a support person (gatekeeper) is required to ensure the effective administration of an LLS. Their prototype supported by a well-defined process map serves as a good model for those companies involved in construction- or engineering-type projects.

MeetingsNet

MeetingsNet is an association that provides a portal for information and resources related to planning meetings and events. MeetingsNet offers a variety of meeting-related magazines discussing planning issues, trends, and events in five focused areas: association, corporate, financial and insurance, religious, and medical. They also produce a weekly e-newsletter. They are strategically connected to the American Society of Association Executives (ASAE). Meetings as a topic for LL make for an interesting review. In MeetingsNet article entitled Lessons Learned from Going Hybrid: Three Case Studies, the discussion focused incorporating hybrid vehicles into a meeting event. While this task might seem simple in nature, it appeared to be more challenging than anticipated, which resulted in an interesting LL.

The format for the LL report was informal, but provided a detailed account of meeting participation, which include 2,656 people. The use of descriptive statistics enabled the reader to grasp metrics pertaining to the attendees, e.g., there were actually 3,479 registrants, but only a 76-percent show rate. The LL write-up was segmented by key topics: speaker management, time, use of streaming technology, and exhibitors.

This format for LL could be classified as key learnings or take-aways. For example, it was mentioned that streaming technology was very expensive; $33,000 to feed four concurrent sessions. There were innovative concepts implemented (contrary to upper management buy-in), such as the use of 3D and virtual reality incorporating avatars. Another key learning was that some presenters may not want their presentations uploaded to a virtual audience. There was a candid response to the level of complexity involved in delivering a hybrid meeting.

NASA

NASA's LL database is regarded as one of the largest and best-designed systems worldwide (Oberhettinger, 2005). NASA (1997) explains that LL is knowledge gained through experience; it may be a positive or negative, but must be significant, factual, technically correct, and applicable. NASA's increased focus on LL is due in part to the major catastrophes, such as the Space Shuttle Challenger accident in 1986. It is also the result of the many new innovations in technology that have been introduced as the result of experiments in outer space.

At NASA, the collection and processing of LL is administered at each site. Authors develop recommendations; actively solicit LL material, contractor, and industry sources; review significant events that impact mission success; and validate LL with subject matter experts. NASA has employed an infusion process to close the loop on actionable LL recommendations at the center and headquarter levels and has designed an integrated communication system, which consists of e-mail, workshops, and newsletters to inform employees of the LL process flow. The Web site for NASA's LLS is available to the public.

NASA's approach to LL, while technologically sound, has been determined to lack the desired expanded capability of just-in-time information distribution to other government agencies. In the Report to the Subcommittee on Space and Aeronautics, Committee on Science, House of Representatives, NASA stated the need to strengthen LL in the context of its overall effort to develop and implement an effective knowledge management system. Recommendations include improvements in strategic planning, coordination, culture, and system enhancements. NASA must continue to optimize technology and leverage integration concepts of LL.

Raytheon Space and Airborne Systems

Raytheon presented a comprehensive process defined as Criteria for LL based on Capability Maturity Model Integration (CMMI), which is a process improvement approach that helps organizations improve performance. Raytheon outlined several key success factors for LL:

- Define terms
- Strategic plan
- Define an LL process
- Target performance measures

Raytheon stressed the concept of garbage-in-garbage-out by stating that lessons in the database must be relevant, well organized, validated according to established standards, and verified. The purpose of verification is to allow customization of the repository to the standards of the group or department. Lessons should address an issue to be investigated or provide pertinent information. A lesson is not a restatement of a policy or process. Accountability for lessons is placed on management to ensure that lessons are carried out. Lessons are checked and monitored by the gatekeepers. According to Raytheon, a lesson must be:

- Significant
- Valid
- Applicable

Raytheon's system was designed to support continuous learning through an LL concept referred to as project retrospectives and after-action reviews. The lessons were pushed out to users. Additionally, forums were held to support LL discussions.

University of Massachusetts

The University of Massachusetts conducted an LL pertaining to workplace safety and health. The article is entitled When My Job Breaks My Back: Shouldering the Burden of Work-related Musculoskeletal Disorders (MSDs) (Markkanen et al., 2001). While the write-up for this LL is an excellent documentary by the university, it paints an ugly picture regarding the transferability of lessons.

The article began with a scenario of a dedicated nurse who had been caring for patients for years and now has traumatic back pain due to servicing them. The extent of her injuries will not allow her to live a normal life. The article went on to say that workplace ergonomic injuries remain one of the most significant occupational health challenges. It provided high-level statistics on MSDs by job type, identifying parts of the body that are subject to MSDs, e.g., lower back, shoulder, neck, elbow/forearm, and carpal tunnel.

Later in the article, attention was shifted to OSHA. A timeline was presented to show the response of OSHA over time with respect to MSD legislation. It said that inspectors have the power to issue citations for ergonomic hazards by invoking the General Duty Clause of the Occupational Safety and Health Act (Section 5(a)): "Each employer shall furnish to each of his employees employment and a place of employment which are free from recognized hazards that are causing or are likely to cause death or serious physical harm to his employees." While it appears that OSHA has been responsive, the level of complexity surrounding MSDs is not just overwhelming, but alarming. The use of LL in this context, of course, is valuable, but may provide only subtle improvements in legislation over time due to the saturation of bureaucracy and politics. This represents a valuable, but ugly situation for LL.

References

Leake, D., Bauer, T., Maguitman, A., & Wilson, D. 2000. Capture, storage and reuse of lessons about information resources: Supporting task-based information search. Retrieved on August 11, 2011 from ftp://ftp.cs.indiana.edu/pub/leake/p-00-02.pdf.

Markkanen, P., D. Kriebel, J. Tickner, and M. Jacobs. 2001. *Lessons learned: Solutions for workplace safety and health. When my job breaks my back: Shouldering the burden of work-related musculoskeletal disorders.* Boston: University of Massachusetts.

Oberhettinger, D. (2005). *Workshop on NPR 7120.6, the NASA lessons learned process: Establishing an effective NASA center process for lessons learned.* Retrieved May 13, 2007 from trs-new.jpl.nasa.gov/dspace/bitstream/2014/37618/1/05-1018.pdf.

Stephens, C., J. Kasher, A. Welsh, Plaskoff. 1999. How to transfer innovations, solutions and lessons learned across product teams: Implementation of a knowledge management system. Proceedings of the 30th Annual Project Management Institute 1999 Seminars and Symposium. Philadelphia, PA: Project Management Institute.

Yin, R. 2008. *Case study research: Design and methods.* Thousand Oaks, CA: Sage Publications.

Lessons That Apply to This Chapter

1. LL can support the merger or acquisition of one company to another.
2. The realization that research is an important part of LL and the more complex the lesson, the more involved the research will be.
3. Large-scale LL systems should allow for transparency throughout the organization to ensure optimization when possible.
4. Web sites that allow public access must carefully scrutinize what lessons are published.
5. In the interest of national security, contribution to defense Web sites should be considered carefully.
6. Regulated companies, e.g., pharmaceuticals, should invest in certifying project managers as well as offer training in evaluation to support competency development.
7. Process mapping can provide a logical insight into determining if/when scenarios that pertain to LL.
8. Having an administrator (gatekeeper) to support LL is essential.
9. Using blended approaches to distribute lessons will ensure more people are reached.
10. Publishing an LL case study on the Internet can be beneficial for those in and outside of the company's industry.

Suggested Reading

Brinkerhoff, R. 2006. *Telling training's story: Evaluation made simple, credible and effective—using the success case method to improve learning and performance*. San Francisco: Berrett-Koehler Publishers.

Chua, A., W. Lam, and S. Majid. 2006. Knowledge reuse in action: The case of CALL. *Journal of Information Science* 32, 251–260.

Cowles, T. 2004. Criteria for lessons learned: A presentation for the 4th annual MMI technology conference and user group, (November 16, 2004). Denver, CO: Raytheon, Inc. Retrieved on August 11, 2011 from www.dtic.mil/ndia/2004cmmi/CMMIT2Tue/LessonsLearnedtc3.pdf.

Goodrum, P., M. Yasin, and D. Hancher. 2003. Lessons learned system for Kentucky transportation projects. Online from www.ktc.uky.edu/Reports/KTC_03_25_SPR_262_03_1F.pdf (accessed July 12, 2007).

Hatch, S. 2011. Lessons learned from going hybrid: Three case studies. Online from http://meetingsnet.com/associationmeetings/news/ hybrid_meeting_case_studies_virtual_edge_0209/ (accessed February 27, 2011).

Kozak-Holland, M. 2002. *On-line, on-time, on-budget: Titanic lessons for the e-business executive* (*Lessons from history* series). Parsippany, NJ: MC Press.

Perrot, P. 2001. Implementing Inspections at AirTouch Celluslar: Verizon Wireless. Retrieved on August 11, 2011 from http://sasqag.org/pastmeetings/Implementing%20Inspections%20at%20AirTouch%20Cellular%202-2001.ppt

The Federal Response to Hurricane Katrina Lessons Learned. February 2006. Online from http://www.whitehouse.gov/reports/katrina-lessons-learned.pdf (accessed January 15, 2008).

Chapter 7

Lessons Learned Scenarios in Real Life

Key Learnings:

- Scenarios support competency development and capacity building.
- Role Plays can emphasize Lessons Learned (LL) in defined situations.
- Address "worst case scenario" thinking through adequate preparation.

Scenario Analysis

A scenario is an event that could possibly occur under particular circumstances. A scenario can be real or make-believe. Scenario development is used in project management to support organizational effectiveness with respect to competency development and capacity building. Scenarios help organizations to outline strategies and develop solutions. Scenario analysis is a process of analyzing possible future situations by considering alternative outcomes. Scenario analysis can be used to uncover "wild cards." For example, analysis of the possibility of a tsunami impacting Hawaii is low, the damage inflicted is so high that the event is much more important (threatening) than the low probability (in any one year) alone would suggest.

Scenarios, like case studies, provide a good training environment for LL. Scenarios can be fictional while at the same time symbolic of real life. Scenarios can be used to develop talking points (scripts) for future conversations with stakeholders. Scenarios can be used in conjunction with logic models, decision trees, or fishbone diagrams to support other analytical processes. This chapter provides several examples to be used as a starting point for scenario development.

Scenarios

Customer Relationship Management

Project

Company Logistics+ has decided to downgrade its supply chain management (SCM) system to support inventory management in the warehouse.

- This will be a stand-alone system.
- They were formerly using a sophisticated e-commerce system from a third-party vendor that went out of business.
- The downgrade is the result of a reduction in products the company is offering.
- This will be the company's primary warehouse, and represents one location.
    - The size of the warehouse is 50,000 square feet.
- Next year the company may decide to open a second warehouse location.
    - It may be located in a different part of the country.
- You have been selected to manage this SCM project because of your overall knowledge of SCM and tenure with the company. You have:
    - Received your CSCP certification from APICS.
    - A working knowledge of project management, but no certification.
- You have been tasked by senior leadership to *capture LL* for this project.

Questions

Who would you go to initially to let the know of the pending request?

What would you initially say to senior leadership who requested your involvement?

Where would you go to address competency development for yourself?

Why would it be important to leverage your APICS certification?

How would you support capacity building for the organization?

When would you let the organization know you would be qualified to conduct LL?

Facilities Management

Project

Government Agency Politics-R-Us has decided to update a training system.

- The new features to the training system will support:
  - Room and location assignment for instructor-led courses
  - E-learning
  - Electronic library
- There will be 4,500 employees tracked in this system.
- The agency may decide to add two modules to this training system next year:
  - Contractor training
  - Policies and procedures documentation
- You are an outside contractor and have been selected to manage this training system project because of your knowledge of training and development. You have:
  - Received your CPLP certification from ASTD
  - Received your CAPM certification from PMI
  - Received a vendor course certificate of completion in Evaluation Practices

You have been contracted by the agency to *implement a LL repository* for this project.

Questions

Who would you go to initially to determine LL system requirements?
What would you initially say to the agency who requested your involvement?
Where would you go to find information on LL repositories?
Why would it be important to research LL repositories?
How would you support the LL repository after implementation?
When would you let the agency know how you plan to proceed?

Professional Association Management

Project

The Association to Improve People (ATIP) will implement a new online meeting registration system.

- ■ This will be the association's central registration system located in:
  - – Chicago
- ■ The registration system will support:
  - – Monthly chapter meetings
  - – Regional conferences
- ■ There will be 1,2000 members tracked in this system:
  - – The association may decide to add another module to this system next year
    - • Meeting evaluations for the purpose of LL
  - – You have been selected to manage this registration system project. You have:
    - • Received your certification from ASAE
    - • A basic understanding of PM in PRINCE2
    - • No knowledge of evaluation principles
- ■ You have been asked by the association to look into conducting LL.

Questions

Who would you go to initially to understand LL?

What would you initially look for in these LL?

Where would you go to generate data or information about LL?

Why would it be important to research LL?

How would you report findings on LL to the association?

When would you look to propose moving forward with LL?

Developing Role Plays Based on Scenarios

The scenarios above were intended to depict actual situations the PE may encounter. In order to adequately prepare for these types of scenarios, it is recommended that the project leader serve as trainer to conduct role plays.

A role play is an outlined discussion between two or more people to increase comfort in the areas of research, measurement and evaluation, which are core, common and critical to PE. A typical role play scenario will involve one PTM acting as the client and the other PTM acting as an evaluator. The intent of the role play is to practice what is going to be said to the client before the PTM meets with them. A rehearsal increases the level of evaluation quality because it is essentially the creation of a script.

There are a variety of situations in which role plays are applicable, such as:

Table 7.1   Made-up Character Names for Role Plays

Annie Body	Betty Kant	Crystal Waters	Doug Hole
Earl Lee Byrd	Faith Hope	Gene Peeples	Harry Arms
Iona Ford	Joe King	Karol Singer	Les Moore
May Flowers	Neil Rizer	Olive Branch	Pete Moss
Quincy Wency	Rose Bush	Sue Yu	Terry Bull
Upton Downs	Virginia Beach	Will Powers	Xavier Schools
Yo Waits	Zach Free	Zero Null	Trinity Tres
Sven Marks	Octavious	Nina Niners	Decimus Tenners

- Participant Observation
- One-on-one Interviews
- Focus Groups

Role plays for the purposes of LL must be:

- Realistic
- Applicable to the environment
- Engaging

Role plays have a tendency to make the participants:

- Feel awkward
- Experience embarrassment
- Show humor

To be sensitive to participant feelings and create an enjoyable learning experience, an artificial person should be created. Using made up names as show in Table 7.1 can be helpful.

Advanced Role Playing

Role playing can be beneficial for experienced project team members. It can assist them in handling challenging situations with stakeholders. Training will most likely be required to familiarize team members with visualization methods such as:

- Process Maps
- Logic Models
- Work Breakdown Structures
- Risk Indexes
- Value Stream Maps
- Influence Diagrams

It is recommended that trainees involved in advanced role playing being required to make a presentation using the selected form of visualization before meeting initially with stakeholders.

Lessons That Apply to This Chapter

1. Large-scale LL systems are very complex.
2. Competency development is important for persons involved in LL.
3. Capacity building is essential for a sponsoring organization.
4. A job description will help define job performance criteria.
5. A Statement of Work (SOW) will clarify job tasks.
6. Certification is a recognized qualifier to be used in candidate selection for projects.
7. Contractors who are hired to work on projects are expected to have more expertise.
8. Research skills must be engaged at the beginning of the project.
9. Senior management should endorse stakeholder communications.
10. Best-case and worst-case scenarios are common ways of thinking, but the middle of the road may prove to be the most practical.

Suggested Reading

Bamberger, M., J. Rugh, and L. Mabry. 2006. *Real world evaluation: Working under budget, time, data, and political constraints.* Thousand Oaks, CA: Sage Publications, Inc.

Bamberger, M., J. Rugh, M. Church, and L. Fort. 2004. Shoestring evaluations under budget, time, and data constraints. Online from www.prel.org/products/pr_/compendium05/BambergerRughChurchFort.pdf (accessed February 6, 2008).

Gaffney, G. 2000. What is a scenario? Usability Techniques series. Online from http://www.infodesign.com.au/ftp/Scenarios.pdf (accessed March 5, 2011).

Hildebrand, C. 2006. On-demand education. *PM Network*, August, 86. Philadelphia, PA: Project Management Institute.

Hynnek, M. 2002. A real life approach to lessons learned. *Project Management Innovations* 2, 5–6.

Chapter 8

Capacity Building through Evaluation

Key Learnings:

- Capacity building
- High-quality Lessons Learned (LL)
- Conducting evaluations
- Evaluation models
- Research methods
- Measurement practices

What Is Capacity Building?

Capacity Building (CB) in the context of LL refers to organizational growth and development in Project Management and Evaluation (PM&E). It is an ongoing process that involves competency development of employees as well as implementing and maturing supporting processes (i.e., project management methodology) and systems (LLR). CB is directed toward enhancing an organization's performance in LL. There must be an appreciation of PM&E for an organization to embrace CB. There are many benefits to CB for an organization:

- Supports efficient and effective decision-making processes
- Streamlines projects, programs, and portfolios

- Improves policies, procedures, and practices
- Reduces operating costs (direct and indirect)
- Enhances metrics reporting (quantitative) as well as qualitative analyses
- Strengthens evaluation use (summative and formative)

How is capacity built within an organization?

- Training in how to evaluate projects
- Conferences to support networking
- Reviewing BP
- Benchmarking
- Partnering with resources, e.g., evaluation consultants
- Tracking the use of LL in audits or inspections (summative)
- Monitoring the application of LL on future projects (formative)

What are the barriers to building capacity within an organization?

- Sponsorship from senior management
- Time allotment for employee training
- Investment in employee development, e.g., attending conferences
- Reporting on CB, relating it to quality initiatives, e.g., balanced scorecard
- Determining the benefit to the organization (e.g., return on investment or payback period)

High-Quality Lessons Learned

Patton (2001) explains that high-quality LL represent principles extrapolated from multiple sources and independently triangulated to increase transferability as cumulative knowledge of working hypotheses that can be adapted and applied to new situations, a form of utilitarian generalizability. Terrell (1999) states that one of the characteristics of a mature PM organization, as documented by Bellcore's research of best in class organizations, is that LL are collected and reviewed during all phases of the project life cycle.

High-quality LL, when implemented correctly within an organization, can make evaluation an iterative process instead of a one-time event. An LL is the documentation of both positive and negative experiences on a project. It must incorporate candid feedback and straightforward communication. The lessons must incorporate Evaluation Knowledge (EK), and must be able to be easily communicated and understood to the target audience. LL cannot be a form of Morse code where the message is a bleep at a time.

Approaches to Conducting Evaluations

Evaluation is about proving, judging, assessing, analyzing, criticizing, questioning, and a whole host of related activities to make a determination of merit (quality), worth (value), or significance (importance). Hence, there are a wide variety of approaches that support evaluation. The following list is not intended to be all-inclusive, but it lists some of the popular approaches and methods.

- **Audit:** A methodical examination of an evaluand that relies on the use of an instrument, i.e., checklist and purposes to ascertain a pass/fail condition. Audits use LL to set the stage of inquiry by defining a sensitivity level, which is based on previous experience.
- **Action Research:** Inquiry and problem solving to understand root causes. Interventions are enacted to support improvement. Action research relies on LL to support conclusions.
- **Critical Incident Technique (CIT):** This is a set of procedures used for collecting observations of human behavior that have significance. These observations are used to solve practical problems or develop strategies. A critical incident is an event or action that has a recognizable impact, positive or negative. The cornerstone of CIT is LL.
- **Deficiency Assessment:** A gap analysis that identifies met/unmet needs. LL are incorporated to support rationale for process improvement.
- **Findings:** Item noted during an investigation and, therefore, is the out come of an investigation. Findings serve as an input to LL.
- **Gap Analysis:** A method to determine the current state from the desired state or proposed state. What is missing is commonly based on comparative analysis or LL.
- **Investigation:** Fact-finding to determine the honesty, integrity, validity, truthfulness, or accuracy of the evaluand. Investigations are reliant on LL to support trend analysis.
- **Key Performance Indicators (KPIs):** Quantify objectives to reflect strategic performance against goals. Because KPIs frequently involve an analysis of milestones and LL, KPIs are inherently a tool of evaluation. There are four types of KPIs:
  - Quantitative indicators—presented as a number
  - Practical indicators—interface with existing company processes
  - Directional indicators—specify whether an organization is getting better or worse
  - Actionable indicators—are within an organization's ability to effect change
- **Key Success Factors (KSFs):** Also known as (AKA) KPIs.

- **Metaevaluation*:** Coined by Michael Scriven in 1969, this is an evaluation of an evaluation. A metaevaluation can use different criteria or values (usually contingent upon sponsor approval) for the second-level evaluation. The Key Evaluation Checklist (KEC) is the preferred instrument to support metaevaluation. LL can be addressed at the end of the metaevaluation to uncover additional findings.
- **Meta-Analysis*:** This is a method of utilizing evaluations from multiple projects to draw conclusions. The combined results can sometimes produce trends and thereby a more accurate conclusion than can be provided by any single evaluation.
- **Needs Assessment:** AKA Preevaluation is an exploration to determine wants and needs of a project and to prioritize these requests. It may become a feasibility study to determine the need for an evaluation or framework to support a project charter. LL are used as an input in the needs assessment process.
- **Observation:** AKA findings.
- **PERT (Project Evaluation Review Technique):** Analyze project completion by diagramming the tasks, resources, and time. Pessimistic, optimistic, and most likely estimates. The critical path indicates the longest duration anticipated.
- **Reevaluation*:** A secondary review of an evaluation primarily for the purpose of verification. During reevaluation, the same process, values, or criteria are adhered to. The primary evaluator may reflect upon LL as they reanalyze the results of the evaluation.
- **Risk Assessment:** Review of factors that impact project time, cost, scope, or resources.
- **Root Cause Analysis:** Underlying source of the problem and contributing factors.
- **Subevaluation:** A smaller part of a larger evaluation is conducted due to project complexity or other need to handle the evaluation component separately. LL are integrally important to avoid redundancy.
- **SWOT Analysis:** Evaluate strengths, weaknesses, opportunities, and threats involved in a project. It involves specifying the objective of the project and identifying the internal and external factors that are favorable and unfavorable to achieve that objective.
- **Triangulation*:** An approach using two or more methods of evaluation to verify the same conclusion. A needs assessment, metaevaluation, and gap analysis could be used in conjunction to support the development of a project charter.

(*Note:* The asterisk (*) denotes a second-level evaluation process.)

Popularized Evaluation Models

- **ADDIE (for training and development): Analysis**—identify training issues, goals, and objectives; audience's needs; current knowledge training environment and mediums. **Design**—specify training purpose, objectives, create storyboards, design prototypes, design graphics, and content. **Development**—create content and learning materials. **Implementation**—implement training project plan and train-the-trainer. Materials are distributed to participants. **Evaluation**—after delivery, training is evaluated.

- **Blooms Taxonomy:** In 1956, Benjamin Bloom developed a learning taxonomy involving: (1) **Knowledge,** (2) **Comprehension,** (3) **Application,** (4) **Analysis,** (5) **Synthesis,** and (6) **Evaluation.** Lorin Anderson updated the taxonomy as follows: (1) **Remember:** define, duplicate, list, memorize, recall, repeat, reproduce, and state; (2) **Understand:** classify, describe, discuss, explain, identify, locate, recognize, report, select, translate, and paraphrase; (3) **Apply:** choose, demonstrate, dramatize, employ, illustrate, interpret, operate, schedule, sketch, solve, use, and write; (4) **Analyze:** appraise, compare, contrast, criticize, differentiate, discriminate, distinguish, examine, experiment, question, and test; (5) **Evaluate:** appraise, argue, defend, judge, select, support, value, and evaluate; (6) **Create:** assemble, construct, create, design, develop, formulate, and write.

- **DMADV (for project management): Define**—design goals consistent with customer demands and enterprise strategy. **Measure**—characteristics Critical to Quality, product capabilities, production process capability, and risks. **Analyze**—to develop and design alternatives, create a high-level design, and evaluate design capability to select the best design. **Design**—details, optimize design, and plan for design verification. **Verify**—design, set up pilots, implement production process, and transition it to process owners.

- **DMAIC: Define**—problem, customer voice, and project goals. **Measure**—key aspects of current process and collect relevant data. **Analyze** data to determine cause-and-effect relationships. Review relationships and ensure all factors have been considered. Perform root cause investigation. **Improve**—current process based on analysis using techniques such as Design of Experiments (DOE), poka yoke or mistake proofing, and standard work to create a new, future state process. Set up pilots to establish process capability. **Control** future state process to ensure any deviations from target are corrected before they result in defects. Implement control systems, i.e., statistical process control, production boards, and visual workplaces, and continuously monitor process.

- **Dopler Decision Matrix:** Plots expected performance against a continuum of delivery options ranging from self-study to skills training (on-the-job). The range of acceptability is a zone where the method of delivery matches the level of knowledge and skill required. Outside the zone represents inefficiencies or ineffectiveness. These levels of performance are reviewed in the context of a Real World Situation (RWS). **Awareness**—Knowing specific facts and concepts and how they could be applied in a RWS. **Understanding**—Ability to perform with cases that are simpler than what will be encountered in a RWS. **Skill**—Ability to perform competently in a RWS under a variety of conditions and meet performance standards. **Efficiency**—Looks at the least cost approach with respect to cost, time, or resources, and may be willing to accept compromises with respect to knowledge transfer under situation-specific circumstances. **Effectiveness**—Promotes quality as the primary driver with respect to knowledge transfer, but does not necessarily consider practicality, i.e., cost, time, or resources.
- **Kirkpatrick (for training and development): Level 1, Reaction Evaluation**—trainee's reaction to training, i.e., feedback forms; **Level 2, Learning Evaluation**—measurement of increase in knowledge, before and after, e.g., test; **Level 3, Behavior Evaluation**—extent of applied learning on the job (transfer). Trainee is now correctly performing task. **Level 4, Results Evaluation**—impact on environment by trainee. As a result of training, level of productivity increased by xx percent. **Level 5, Return on Investment Evaluation**—payback (cost benefit analysis) as a result of training.
- **PLAN–DO–CHECK–ACT (PDCA): Processes** are developed after objectives are established to deliver results in accordance with the expected output. **Deliver** new processes for implementation. Often on a small scale, if possible. **Compare** actual results to expected results. **Analyze** differences to determine causes.

Research Methods

Research involves a systematic method to obtain facts, figures, and supporting content. If/then hypotheses are an inherent part of research methodology. Approaches to research are quantitative (numbers-oriented) and/or qualitative (values-based). A research protocol is the procedure used to gather information. It should be handled as a subproject and outline the rationale, objectives, methods, populations, time frame, and expected outcome of the research.

Table 8.1   Research Methods by Process Group/Knowledge Areas

	Initiating	Planning	Executing	Monitoring / Controlling	Closing
Communications	Determine Stakeholder Relationships	Focus Groups Interviews Surveys	Participant Observation	Information Distribution	Performance Reporting
Cost	Opportunity / Sunk Costs	Top Down / Bottom Up / Parametric	IRR / NPV	Earned Value Analysis	Payback Period / Return on Investment
Human Resources	Job Role Determination	Job Description	Responsibility Assignment Matrix	Productivity Per Person	Job Evaluation
Integration	Case Studies	Comparative / Competitive Analysis	Internal Benchmarking	Best Practices	Archive Records
Procurement	RFI	RFP	SOW	Service Level Agreement	Contract Signoff
Quality	Criteria / Values	Change Management	Verification / Validation / Testing	Change Control	Inspection / Audit
Risk	Flow Chart/Process Map	Decision Tree / Logic Model	Risk Breakdown Structure	Monte Carlo / Simulation / Delphi	Cause / Effect Analysis
Scope	Requirements Gathering	WBS / WBS Dictionary	Requirements Management	Configuration Management	Value Stream Map
Time	Timeline / Schedule	PERT / AON / AOA	Fast-tracking / Crashing	Critical Path / Dependencies	GANNT / Milestones

Project Team Members (PTMs) spend approximately 90 percent of their time communicating and engaging in a variety of activities to complete research including phone calls, e-mails, meeting, and other forms of correspondence. Research is integrally related to project management and may utilize evaluation-related processes as a framework (Table 8.1).

Measurement Practices

Measurement practices are the assumptions, constraints, theories, approaches, and scales an organization adopts and uses in order to support valid and reliable findings. Validity is concerned with accuracy, correctness, truthfulness, whereas reliability is concerned with consistency, reproducibility, and redundancy. Measurement practices must support both valid and reliable.

What factors determine if a measurement practice is both valid and reliable?

1. Authorities: Who in terms of person or organization indicates it is so?
2. Experience: What empirical evidence supports the hypothesis?
3. Proof: Where in history can we find situations that indicate this approach?
4. Time frame: When is it considered applicable, knowing that things change over time?

5. Rationale: Why is it considered to be so?
6. Justification: How were the conclusions arrived at?
7. Standards: Which guidelines or rules substantiate it?

What are some good examples of measurement practices that are both valid and reliable?

- Project Management Body of Knowledge (PMBOK) published by PMI PRINCE2
- OMBOK published by APICS
- ISO Standards: ISO has developed over 18,500 International Standards
- IEEE Standards
- OSHA Regulations
- FDA Compliance Guidelines
- ANSI Standards

Measurement can help organizations determine effectiveness and efficiency:

- **Effectiveness:** "Best use" of resources to achieve outcome, i.e., performance in how well, how strong, or how good.
- **Efficiency:** "Least cost" of a preferred solution, i.e., saving, minimizing, or streamlining.

Measurement practices include a host of activities that focus on a variety of end points:

- **Benchmarking:** Comparative analysis and competitive analysis
- **Estimating:** Budgeting (e.g., EVM (Earned Value Management)) and Calculating (e.g., ROI)
- **Predicting:** Forecasting and trending

While measurement relies heavily on the use of descriptive statistics or number crunching, it is more than just determining:

- Mean: Summed scores divided by number of scores
- Mode: Most frequently occurring score
- Median: 50 percent of cases in distribution fall below/above
- Ratio: A way of comparing two quantities by dividing them, e.g., 2 to 3
- Percentage: A proportion in relation to a whole, e.g., 25 percent
- Sample Size: Number of observations that constitute it

Lessons That Apply to This Chapter

1. Capacity building is essential to maturing an organization in PM&E.
2. Competency development for PTMs must happen before or in conjunction with organizational capacity building.
3. Organizations should strive for high-quality LL.
4. A needs assessment is essential before any project is undertaken.
5. Metaevaluation should be considered as part of the normal evaluation process.
6. PERT or GANNT charts can be used to evaluate project status.
7. ADDIE combines both a method for development and evaluation.
8. Research protocol should be signed off by senior management.
9. Effectiveness and efficiency are both important to measurement practices.
10. The Kirkpatrick model provides a solid framework for training evaluation.

Suggested Reading

Bangert-Downs, R. 1995. Misunderstanding meta-analysis. *Evaluation Health Profession* 18, 304–315.

Belshaw, B., S. Citrin, and D. Stewart. 2001 *A strong base.* Alexandria, VA: APICS, The Performance Advantage (Nov./Dec.) 54–57.

Bickman, L. 1994. An optimistic view of evaluation. *American Journal of Evaluation* 12, 255–259.

Butler, J. 2005. Metaevaluation and implications for program improvement. Online from http://www.acetinc.com/Newsletters/Issue%2010.pdf (accessed July 6, 2007).

Chou, S., and M. He. 2004. Knowledge management: The distinctive roles of knowledge assets in facilitating knowledge. *Journal of Information Science* 30, 146–164.

Chou, T., P. Chang, C. Tsai, and Y. Cheng. 2005. Internal learning climate, knowledge management process and perceived knowledge management satisfaction. *Journal of Information Science* 31, 283–296.

Cook, T., and C. Gruder. 1978. Metaevaluation research. *Evaluation Review* 2, 5–51.

Cook, T., J. Levinson-Rose, and W. Pollard. 1980. The misutilization of evaluation research: Some pitfalls of definition. *Science Communication* 1, 477–498.

Cooksy, L., and V. Caracelli. 2005. Quality, context and use: Issues in achieving the goals of metaevaluation. *American Journal of Evaluation* 26, 31–42.

Forss, K., B. Cracknell, and K. Samset. 1994. Can evaluation help an organization to learn? *Evaluation Review* 18, 574–591.

Friedman, V., R. Lipshitz, and M. Popper. 2005. The mystification of organizational learning. *Journal of Management Inquiry* 14, 19–30.

Gajda, R. 2004. Utilizing collaboration theory to evaluate strategic alliances. *American Journal of Evaluation* 25, 65–77.

Hertzum, M., and N. E. Jacobsen. 2001. The evaluator effect: A chilling fact about usability evaluation methods. *International Journal of Human-Computer Interaction* 13 (4): 421–443.

Hummel, B. 2003. Metaevaluation: An online resource. Online from http://www.bhummel.com/Metaevaluation/resources.html (accessed August 9, 2007).

Ipe, M. 2003. Knowledge sharing in organization: A conceptual framework. *Human Resource Development Review* 2, 337–359.

Joint Committee on Standards for Educational Evaluation. 1988. *The personnel evaluation standards: How to assess evaluations of educational programs.* Thousand Oaks, CA: Sage Publications, Inc.

Joint Committee on Standards for Educational Evaluation. 1994. *The program evaluation standards: How to assess evaluations of educational programs*, 2nd ed. Thousand Oaks, CA: Sage Publications, Inc.

Kalseth, K., and S. Cummings. 2001. Knowledge management: Development strategy or business strategy. *Information Development* 17, 163–172.

Kang, K., and C. Yi. 2000. *A design of the metaevaluation model.* Taejon, Korea: Chungnam National University.

Kusek, J., and R. Rist. 2004. *Ten steps to a results-based monitoring and evaluation system.* Washington, D.C.: World Bank Publications.

Lawrenz, F., and D. Huffman. 2003. How can multi-site evaluations be participatory? *American Journal of Evaluation* 24, 471–482.

Love, A. 2001. The future of evaluation: Catching rocks with cauldrons. *American Journal of Evaluation* 22, 437–444.

Mark, M. 2001. Evaluation's future: Furor, futile, or fertile? *American Journal of Evaluation* 22, 457–479.

Nelson, S. 1979. Knowledge creation: An overview. *Science Communication* 1, 123–149.

Nilsson, N., and D. Hogben. 1983. Metaevaluation. *New Directions for Program Evaluation* 19, 83–97.

Payne, D. 1988. How I learned to love the standards. *American Journal of Evaluation* 9, 37–44.

Preskill, H. and D. Russ-Eft. 2005. *Building evaluation capacity: 72 activities for teaching and training.* Thousand Oaks, CA: Sage Publications, Inc.

Reeve, J., and D. Peerbhoy. 2007. Evaluating the evaluation: Understanding the utility and limitations of evaluation as a tool for organizational learning. *Health Education Journal* 66, 120–131.

Reineke, R. 1991. Stakeholder involvement in evaluation: Suggestions for practice. *American Journal of Evaluation* 12, 39–44.

Reineke, R., and W. Welch. 1986. Client centered metaevaluation. *American Journal of Evaluation* 7, 16–24.

Rodriguez-Campos, L. 2005. *Collaborative evaluations: A step-by-step model for the evaluator.* Tamarac, FL: Llumina Press.

Schwandt, T. 1989. The politics of verifying trustworthiness in evaluation auditing. *American Journal of Evaluation* 10, 33–40.

Schwandt, T., and E. Halpern. 1990. Linking auditing and metaevaluation: Enhancing quality. *Applied Research* 10, 237–241.

Scriven, M. 1981. The good news and the bad news about product evaluation. *American Journal of Evaluation* 2, 278–282.

Stufflebeam, D. 2001. The metaevaluation imperative. *American Journal of Evaluation* 22, 183–209.

Taut, S. 2007. Studying self-evaluation capacity building in a large international development organization. *American Journal of Evaluation* 28, 45–59.

Torraco, R. 2000. A theory of knowledge management. *Advances in Developing Human Resources* 2, 38–62.

Torres, R., and H. Preskill. 2001. Evaluation and organizational learning: Past, present and future. *American Journal of Evaluation* 22, 387–395.

Tywoniak, S. 2007. Knowledge in four deformation dimensions. *Organization* 14, 53–76.

Ward, T. 2007. The blind men and the elephant: Making sense of knowledge management. Paper presented at the American Evaluation Association 2007 International Conference, Baltimore, MD.

Wiles, P. 2004. Meta-evaluation. Online from http://www.odi.org.uk/ALNAP/publications/RHA2003/pdfs/FCh401bp.pdf (accessed June 15, 2007).

Woodside, A., and M. Sakai. 2003. Meta-evaluation: Assessing alternative methods of performance evaluation and audits of planned and implemented marketing strategies. Online from http://www2.bc.edu/~woodsiar/publications/ ABM&P%20 Sakai.pdf (accessed August 24, 2007).

Worthen, B. 2001. Whither evaluation? That all depends. *American Journal of Evaluation* 22, 409–418.

Yang, C., and L. Chen. 2007. Can organizational knowledge capabilities affect knowledge sharing behavior? *Journal of Information Science* 33, 95–109.

Yates-Mercer, P., and D. Bawden. 2002. Managing the paradox: The valuation of knowledge and knowledge management. *Journal of Information Science* 28, 19–29.

Chapter 9

Project Evaluation Management

Key Learnings:

- Project Management and Leadership
- Evaluation Reporting
- Charts: PERT and Gantt
- Collaborative Evaluation
- Communication Preferences
- Key Evaluation Checklist
- Documentation Standards
- Computers as Evaluators
- Macro versus Microevaluation

Project Evaluation Management

PEM includes the theories, methods, processes, approaches, and work to be performed to manage the determination of merit (quality), worth (value), or significance (importance) in projects. With PEM, there will be a coordinated work effort and quality supporting documentation. Project Management (PM) contains some processes that involve evaluation:

- Cost Analysis
- Job Performance Review

- Procurement Audit
- Risk Assessment
- Vendor Rating

PEM should always consider the second-level evaluation processes, i.e., metaevaluation, reevaluation, triangulation, or metaanalysis. For example, an assessment of the policies used for vendor evaluations would be second-level evaluation processes, which check or double-check the integrity of the evaluation to ensure conformance.

Project Evaluation Plan

Evaluation Plan: A proposal that describes the process that will be used to guide an evaluation. It includes an overview of the evaluand, scope of the evaluation, purpose of the evaluation, supporting research, evaluative criteria, evaluation methodology, timeline, budget, target population, stakeholders, evaluation team, values, evaluative conclusions, recommendations, and references.

Evaluation Reporting

The purpose of evaluation reporting is threefold: convey information, facilitate understanding, and create meaning and support decision making (Torres, Preskill, and Piontek, 2005). The Board of Regents of the University of Wisconsin System say there are four basic elements of good evaluation reporting:

1. Consider the needs of your target audience(s) even before you begin your evaluation.
2. Give your audience important details of the evaluation.
3. Use caution in reporting findings and drawing conclusions.
4. Have others read or listen to your report and give you feedback before you create the final version.

Good Documentation Practices versus Documentation Quality Standards

Documents are objective evidence that tasks or activities have been performed. The saying goes: "If it is not documented, it didn't happen." Good Documentation Practices (GDP) describes standards by which documents are created and maintained and demonstrate conformance to requirements. It is the program that supports the writing of standard operating procedures (SOPs) and supporting training documents. GDP are expected of regulated industries, such as pharmaceutical companies.

Documentation Quality Standards (DQS) describes the policies, practices, and procedures by which documents are created and maintained. DQS is a primary function of PEM that is intended to ensure that documents conform to an organization's quality standards. DQS is very similar to GDP, and is universal to provide a framework for nonregulated industries (Table 4.)

Ten universal DQS that apply to PEM documentation include:

1. **Legible:** Letters and numbers are clear and readable to avoid misinterpretation.
2. **Understandable:** Language, meaning, terms, and definitions are understood by readers.
3. **Organized:** Labeled, numbered, and paginated so information can be followed.
4. **Accessible:** Retrieved and viewed by authorized persons upon reasonable request.
5. **Truthful:** Factual and accurate to the best of the contributor's knowledge.
6. **Ethical:** Honest and legal to the best of the contributor's knowledge.
7. **Permanent:** Automatically dated when created and saved.
8. **Consistent:** Incorporate quality attributes through an agreed to common appearance.
9. **Controlled:** Have an established workflow to allow for review and approval.
10. **Reviewed:** Checked by self or others to ensure content meets requirements.

Visualization and Presentation

Visualization and Presentation (V&P) is an overarching concept that addresses techniques for delivery of Lessons Learned (LL). Visualization is a technique for creating images to communicate a message. A visual image is a mental image that is based on visual perception. People will perceive things differently, so common ground must be established, which is why there is a need for presentation. Presentation is the practice of showing, explaining, or demonstrating the content of a topic to an audience. Visualization is mental and presentation is physical. At the end of the project, during a final review of LL, a project dashboard can be used. A dashboard can be real time or static, and present information in a way that is descriptive and intuitive to the viewer.

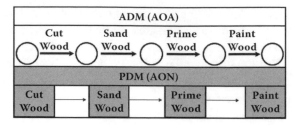

Figure 9.1 Arrow versus precedence diagramming methods.

Project Management Network Diagrams

A Project Management Network Diagram (PMND) is a visual display of a project schedule. A popular complement to a PMND is a PERT or Gantt chart. There are two types of network diagrams: Arrow Diagram Method (ADM) and Precedence Diagram Method (PDM). ADM puts tasks on arrows, which are connected to dependent activities with nodes. This method is also referred to as Activity on Arrow (AOA). Conversely, PDM depicts nodes for events and arrows for activities. PDM is also called Activity on Node (AON) (Figure 9.1).

PERT

One of the most popular forms of PEM is the Project (or Program) Evaluation Review Technique (PERT), which is used to analyze and represent the tasks involved in completing a project. PERT provides the pessimistic, optimistic, and most likely time estimates for a project. PERT was created to simplify the planning and scheduling of large and complex projects. PERT was initially developed in 1957 for the U.S. Navy Special Projects Office to support the U.S. Navy's Polaris nuclear submarine project. PERT is commonly used in conjunction with the Critical Path Method (CPM). CPM calculates the longest path of planned tasks to the end of the project, and the earliest and latest that each task can start and finish without making the project duration longer. If two or more durations are of equal length, then two or more critical paths may exist.

Gantt

Another popular method to support PEM is the Gantt chart. Since its inception in 1910, Gantt charts (Gantts) illustrate the start and finish dates, dependencies (mandatory and discretionary), and key elements that comprise the Work Breakdown Structure (WBS) during the Project Life Cycle (PLC). Gantts show current schedule status using percentage-complete. Gantts are a common technique for representing the PLC and milestones. Gantts represent cost, time, and scope of projects, and focus primarily on schedule management. Therefore, the evaluation becomes centrally focused on time management as it pertains to resources (i.e., people, systems, facilities, equipment, and materials). A Gantt is evaluative in nature because it supports the determination of merit, worth, or significance through the graphic illustration of the project schedule. In Microsoft Project®, CPM can be displayed by using the Gantt Chart Wizard. Gantts can become somewhat unwieldy for projects with a large number of activities, requiring wider computer displays.

PEM as a New Knowledge Area?

Process Inputs, Tools, Techniques, and Outputs (ITTO) represent a special category of information in the Project Management Body of Knowledge (PMBOK). ITTO referenced in the PMBOK include primary items. It would not be practical to list every conceivable ITTO. Recommended changes to the PMBOK for ITTOs or other aspects are submitted by industry experts during the review cycle. If there could be a new tenth knowledge area represented in the PMBOK, PEM would be a good candidate. The model in Figure 9.2 presents a similar look and feel to the other nine knowledge areas in the PMBOK, which are presented in terms of ITTO.

- **Input:** Items or actions (internal or external) required by a process before it can proceed. It may involve different resources (people, facilities, equipment, materials, or supplies).

Figure 9.2 Conceptual project evaluation management as a Knowledge Area (KA).

- **Tool:** Something tangible, such as a template used to generate a product, service, or result. Templates include checklists, forms, guidelines, models, standards, etc.
- **Technique:** An approach, method, or theory used to complete an activity. For example, a GBP can be used to specify procedures.
- **Output:** A product, service, or result that is realized by a process. An outcome is a form of output that may be used to determine impacts.

Collaborative Evaluation

Collaborative Evaluation (CE) is a partnership approach to conducting LL that engages stakeholders in the evaluation process. By contrast, in non-CE evaluation approaches, subjects are observed and not allowed to participate in the evaluation process. While CE is supported by evaluation theorists and practitioners and has found to be very effective, there is some concern amongst evaluators that limitations should be imposed.

- Evaluation processes must be established and adhered to and not impacted by stakeholders who feel empowered to make or alter decisions.
- Standards for evaluators must be set and competencies of those involved in the evaluation should meet those criteria.

Microevaluation versus Macroevaluation

Microevaluation or macroevaluation, that is the question! In microevaluation, each component of the project is analyzed: the five PGs and nine KAs. As a result, performance from the beginning to the end of the project can be realized. Alternatively, demands on the triple-constraint can be determined, i.e., cost, time, and scope. In macroevaluation, the whole project is judged. Considerations in macroevaluation may involve opportunity costs, sunk costs, payback period, and return on investment (ROI). So, the answer to the question above, in fact, is both micro and macro approaches should be considered a part of PEM. Programs such as MS Project can be very effective tools to support either approach.

Key Evaluation Checklist

The Key Evaluation Checklist (KEC) is a well known standard for PEM. It provides a step-by-step process. It can also be used as a template for an evaluation proposal or evaluation report. The KEC is summarized below.

KEC (for general evaluation): Preliminaries (Part A) includes **Executive Summary,** summarize the results and investigatory process; **Preface,** define the client, audience, stakeholders, and other impactees; **Methodology,** how will the evaluation be handled in terms of tools and techniques. Foundations (Part B) includes: (1) **Background and Context,** identify historical, recent, concurrent, and projected settings for the program; (2) **Descriptions and Definitions,** record official descriptions of program and components, context, and environment. (3) **Consumers,** stakeholders; (4) **Resources,** (SWOT analysis) of people, equipment, facilities, and materials; (5) **Values,** criteria to be adhered to. Subevaluations (Part C) includes (6) **Process,** this is the assessment of the quality, value, or importance of everything that happens or applies before true outcomes emerge; (7) **Outcomes,** impact on stakeholders; (8) **Costs,** investments required; (9) **Comparisons,** may involve benchmarking or best practices; (10) **Generalizability,** is transferable to other situations. **Conclusions and Implications** (Part D): (11) **Synthesis,** the basics of Part B are combined with the subevaluations of Part C, including synthesis of empirical results and values into the overall evaluation; (12) **Recommendations and Explanations,** the use of LL emerges during this phase; (13) **Responsibility and Justification,** accountability of results; (14) **Report and Support** for evaluative conclusions; (15) **Metaevaluation,** an evaluation of the evaluation should be conducted.

Reference

Torres, R. T., H.S. Preskill, and M. E. Piontek. 2005. *Evaluation strategies for communicating and reporting: Enhancing learning in organizations,* 2nd ed. Thousand Oaks, CA: Sage Publications.

Lessons That Apply to This Chapter

1. PEM must be perceived as a necessary function to support active team engagement.
2. PERT is an excellent model to support PEM because it is graphical in nature.
3. Second-level evaluation is an integral part of PEM.
4. There can be multiple critical paths on a project if the longest duration is composed of two paths that are equal.
5. Gantt charts have gained universal acceptance and are easy to understand.

6. Visualization is a powerful tool that can enhance content and make it more understandable.
7. Each organization should define ten key characteristics for DQS or GDP.
8. DQSs are important for businesses that value sound records management practices.
9. GDP are essential in maintaining good relations with the FDA.
10. MS Project is the most popular project management software.

Suggested Reading

Avison, D., and G. Torkzadeh. 2009. *Information systems project management.* Thousand Oaks, CA: Sage Publications, Inc.

Board of Regents. 2002. *Basics of good evaluation reporting.* Madison, WI: University of Wisconsin-Madison.

Brown, S. 2000. *Customer relationship management.* Etobicoke, Ontario: John Wiley & Sons.

Hale, J. 2002. *Performance-based evaluation: Tools and techniques to measure the impact of training.* San Francisco: Josey-Bass Pfeiffer.

Kerzner, H. 2004. *Project management: A system approach to planning, scheduling and controlling.* Hoboken, NJ: John Wiley & Sons.

Kerzner, H. 2009. *Advanced project management: Best practices on implementation.* Hoboken, NJ: John Wiley & Sons.

Kerzner, H., and F. Saladis. 2009. *Value-driven project management.* New York: IIL Publishing.

Kusek, J., and R. Rist. 2004. *Ten steps to a results-based monitoring and evaluation system.* Washington, D.C.: World Bank Publications.

Lewkowicz, J., and D. Nunan. 1999. *The limits of collaborative evaluation.* Hong Kong, China: University of Hong Kong.

Project Management Institute. 2006a. *Practice standard for work breakdown structures,* 2nd ed. Philadelphia: Project Management Institute.

Project Management Institute. 2006b. *Practice standard for configuration management.* Philadelphia: Project Management Institute.

Rad, P., and G. Levin. 2006. *Project portfolio management tools and techniques.* New York: IIL Publishing.

Rodriguez-Campos, L. 2005. *Collaborative evaluations: A step-by-step model for the evaluator.* Tamarac, FL: Llumina Press.

Witkin, B., and J. Altschuld. 1995. *Planning and conducting needs assessment.* Thousand Oaks, CA: Sage Publications, Inc.

Chapter 10

Project Evaluation Resource Kit (PERK)

Key Learnings:

- Organizing Project Evaluation Documentation
- PERK Overview
- Supporting Texts
- Software Needed
- System Requirements and Updates

Keeping Project Evaluation Documentation Organized

When information is not raining in projects, it is snowing. There is literally a blizzard of content Project Team Members (PTMs) are required to track to ensure project success, e.g., risk assessments, earned value, and work packages. Selecting the right tools will make the work for PTMs easier. There are many good options to choose from. Some nifty Web-based applications, such as Google Docs, Basecamp, Dropbox, Dimdim, and Mindmeister support information management on the computer. Using PERK with these types of applications represents Business Practices (BP).

There are other times, where hard-copy printouts are preferred and solutions also should be developed to support these requirements. Companies, such as SMEAD, produce project jackets, which are a good way to keep project materials organized. There are preprinted areas on the front, back, and tab to provide a means for tracking projects. If hanging files are to be used, then folders within folders can provide improved filing of project information (Allen, 2007).

Systematizing the Temporary and Unique

Project Management (PM) is considered to be temporary and unique. However, the systems that support PM should not be, especially when it comes to evaluation. Organizations are relying more and more on software to perform evaluation-related tasks. Microsoft® Project has been available since 1984. It has helped project managers develop plans, assign resources, manage budgets, track progress, and analyze workloads. While MS Project is great, there are other programs to support PM. For example, Lewins and Silver (2007) discuss using computers to perform qualitative research functions. Strings of words can be linked together to develop content maps for further analysis.

While software is able to perform routine quantitative and sometimes qualitative functions, programs currently lack the ability to compute evaluative conclusions and apply Lessons Learned (LL). For example, a project could be behind schedule and over budget. While the software could estimate what can be done regarding resource allocation, it is not equipped to analyze PTM competencies.

Establishing Communication Preferences

There are a variety of ways PTMs network to support strategies for communicating and reporting evaluations and LL (Torres, Preskill, and Piontek, 2005). The nature of LL frequently requires multiple locations to be engaged at the same time. Synchronous Online Communication (SOC) can effectively support virtual environments, such as:

- **Instant Messaging:** Twitter and Google Chat provide real-time e-mail functionality.
- **Meetings:** Microsoft® NetMeeting (small group) and Microsoft® Office Live Meeting (large group) work well to display contents of the desktop and can use a separate line for voice communication.
- **Web-Conferencing:** Webex (presentation and discussion) and Skype® (virtual classroom).

PERK Overview

The Project Evaluation Resource Kit (PERK) is a comprehensive tool to support Project Evaluation Management (PEM). The purpose of the PERK is to help organize Project Evaluation (PE) documents and keep PTMs "on the same page," i.e., to support LL. While PERK can be used for all types of PE, it is specifically designed for LL. PERK is a companion to this book (*The Basics of Lessons Learning in Projects*). It is provided free of charge and not sold separately. PERK is a SURE BET (Smart Useful Realistic Essential Beneficial Evaluation Templates) and supported at http://www.lessonslearned.info.

PERK is based on BP in PEM. It is not just a book of forms, but rather a heavy duty toolbox of documentation required to successfully evaluate projects and perform the required ongoing maintenance on project lessons. PERK templates includes contract examples, job aids, guidelines, procedures, forms, checklists, report designs, diagrams, presentations, plans, supporting media, and more.

While resources, such as the Project Management Body of Knowledge (PMBOK) and Projects in Controlled Environments (PRINCE2) address PM methods and practices, PERK fills an identified gap by providing supporting resources for PE. PERK is an excellent supplement that enables PTMs to apply the tools and techniques of LL in projects. These documents are logically organized by process group and knowledge area and are fully editable. The contents in PERK are customizable to meet the needs of the organization. The extent to which the templates will be used is dependent upon the project. The PERK includes an e-learning program that walks the PTM through the process of using the material. PERK connects with key industry publications, e.g., PMBOK, which should be referenced when using PERK. It also requires industry standard software to open the templates, i.e., Microsoft® Word.

Top Ten Publications to Support PERK

A comprehensive literature review was conducted while creating PERK. Some topic areas are large in scope and references should be consulted to ensure the desired level of coverage.

1. **The Basics of Lessons Learned in Projects**: This is the key text that provides comprehensive information on how to evaluate projects in the context of LL.
2. **PMBOK**: It is necessary to refer to the primary standard to develop an in-depth understanding of the process groups and knowledge areas and related processes.

3. **PRINCE2**: This framework for PM can utilize PERK with minor adjustments.
4. **OMBOK**: Provides the framework for Supply Chain Management (SCM), which is frequently an integral activity of PM.
5. **Project Manager's Book of Forms**: Templates that apply to PM overall and can supplement required documentation.
6. **Evaluation Strategies for Communicating and Reporting**: Outlines step-by-step methods for completing evaluation reports.
7. **Collaborative Evaluations**: The approach to conducting LL utilizes this philosophy.
8. **Evaluation Thesaurus**: Provides comprehensive terms and definitions on evaluation.
9. **Post Project Reviews to Gain Effective Lessons Learned**: Substantiates the need for LL by providing survey results of 500 organizations.
10. **Information Graphics: A Comprehensive Illustrated Reference**: Definitive reference for creating diagrams to support visualization.

Top Ten Software Aligned to Support PERK

The following programs are required to utilize PERK resources. *Note*: PERK updates will not change the required software list below and additional resources to PERK may be added.

1. **MS Word:** Introductory position letter, Workshop checklist, Brainstorming Index Cards, Testimonial Release Form, Change Control form, Learning Journal, Audio/Video Transcript, Project Charter, Project Plan Outline, Good Business Practices Procedure, Vendor Quality Agreement, Vendor Selection Checklist, Statement of Work, Project Evaluator Job Description, Project Evaluator Performance Evaluation, Project Evaluator Competency Development Plan
2. **MS Excel:** PM Formula Worksheet, Training ROI Calculator, Project Evaluator Responsibility Assignment Matrix
3. **MS Access:** Training Resources and Information Library
4. **MS PowerPoint:** Overview presentation, presentation templates, Stakeholder Relationship Grid
5. **MS SharePoint:** SharePoint Administrator Job Aid
6. **MS Project:** Gantt, Project Schedule
7. **MS Visio:** Work Breakdown Structure, Risk Breakdown Structure, Project Network Diagram
8. **Adobe® Acrobat:** Records retention procedure, Interview Protocol, Project Evaluator Code of Conduct, Project Evaluation Standards, PT Agreement,

Glossary of terms and definitions, Collaborative Learning Map (TELL), Time Management Tips

9. **Adobe Dreamweaver:** Web-based survey
10. **Adobe Premiere:** LL Kickoff Motivational Video

Follow THIS

Tips, Hints, Instructions Sheet (THIS) are directions and supporting information you need to get the most out of PERK. THIS is more than a step-by-step guide, but rather best practices and good ideas.

System Requirements

This CD is compatible with both Windows and Macintosh operating systems. A CD ROM drive and 32 MB RAM is required. All documents are in MS Office format: MS Word, MS Excel, MS PowerPoint, MS Project, and .HTML. Multimedia files are .MP3 and .SWF. Graphic elements are .jpg, .gif. Web components utilize .CSS, .HTM, .HTML.

Accessing the CD

Insert the PERK CD into the CD ROM drive. Click the file PERT.HTML in the root directory to display the CD contents. Alternatively, the CD can be accessed directly by clicking on the associated directory and then the file corresponding to the desired content. For example, access the Integration directory, then locate KAINTE01 to view the LL Project Charter Template. PERK is in digital format only and the documents are not displayed in the book.

Multiuser Access to CD Contents

In a multiuser environment, i.e., SharePoint server, contents for the CD may be loaded to an enterprise server provided that each user who accesses the CD contents purchases the book. Therefore, if ten users will perform LL and use the CD contents, ten copies of the book must be purchased.

References

Allen, D. 2007. *Getting things done: The art of stress-free productivity.* New York: Penguin Press.

Lewins, A. & Silver, C. 2007. *Using Software in Qualitative Research: A Step-by-Step Guide.* Los Angeles, CA: Sage Publications.

Torres, R., H. Preskill, and M. Piontek. 2005. *Evaluation strategies for communicating and reporting*, 2nd ed. Thousand Oaks, CA: Sage Publications, Inc.

Lessons That Apply to This Chapter

1. Using PERK effectively requires reading and understanding this text as a prerequisite.
2. Standardizing project evaluation documentation is a best practice.
3. PERK provides one-stop shopping to organize project evaluation documentation.
4. PERK contents can be customized to meet your specific needs.
5. Maintain documents in an electronic format whenever possible.
6. Before using PERK, complete the e-learning program included in the book.
7. Software to supplement PERK includes MS Office, MS Project, Smartdraw.
8. PERK has been formatted to be compatible with Windows or MAC platforms.
9. Supplements to PERK include the PMBOK, Prince2, and Project Managers book of forms.
10. Legal review of PERK contracts is recommended to ensure compliance with your organization.

Suggested Reading

Bichelmeyer, B. 2003. Checklist for formatting checklists. Online from http://www.wmich.edu/evalctr./checklists/checklistsmenu.htm (accessed September 23, 2007).

Few, S. 2006. *Information dashboard design: The effective visual communication of data.* Sebastopol, CA: O'Reilly Media.

Golden-Biddle, K., and K. Locke. 2007. *Composing qualitative research.* Thousand Oaks, CA: Sage Publications, Inc.

Harris, R. 1999. *Information graphics: A comprehensive illustrated reference.* New York: Oxford University Press.

Harroch, R. 2000. *Business contracts kit for dummies.* New York: John Wiley & Sons.

Knowlton, L., and C. Phillips. 2009. *The logic model guidebook.* Thousand Oaks, CA: Sage Publications, Inc.

Lee, Q., and B. Snyder. 2006. *Value stream and process mapping.* Bellingham, WA: Enna Products Corporation.

Locke, L., W. Spirduso, and S. Silverman. 2007. *Proposals that work.* Thousand Oaks, CA: Sage Publications, Inc.

Stackpole, C. 2009. *A project manager's book of forms.* Philadelphia: Project Management Institute.

Stover, T. 2004. *Microsoft Office project 2003 inside out.* Redmond, WA: Microsoft Publishing.

Stufflebeam, D. 2000. Guidelines for developing evaluation checklists. Online from http://www.wmich.edu/evalctr./checklists/checklistsmenu.htm (accessed September 10, 2007).

Stufflebeam, D. 2001. Evaluation checklists: Practical tools for guiding and judging evaluations. *American Journal of Evaluation* 22, 71–79.

Appendix

Abbreviations

ADM: Arrow Diagram Method
AHP: Analytical Hierarchy Process
AKA: Also Known As
ALT: Adult Learning Theory
APICS: Advancing Productivity Innovation for Competitive Success
BI: Business Intelligence
BP: Best Practices
BPLL: Best Practices Lessons Learned
CB: Capacity Building
CPM: Critical Path Method
DIKUD: Data–Information–Knowledge–Understanding–Decision
DQS: Documentation Quality Standards
EVM: Earned Value Management
FAQ: Frequently Asked Question
FDA: Food and Drug Administration
GBP: Good Business Practice
GDP: Good Documentation Practices
GLLSP: General Lessons Learned Service Provider
GLP: Good Laboratory Practices
GMP: Good Manufacturing Practices
GPB: Good Practice Bulletin
ILM: Information Life Cycle Management
IS: Information Systems
IT: Information Technology
ITTO: Inputs, Tools and Techniques and Outputs
KA: Knowledge Area
KM: Knowledge Management
KPS: Key Project Stakeholder

LL: Lessons Learned
LLR: Lessons Learned Repository
LLRRB: Lessons Learned Repository Recycle Bin
LLSS: Lessons Learned Support System
MS: Microsoft
NEEDS: Necessity, Expertise, Economics, Documentation, and Systems
NHTSA: National Highway Traffic Safety Administration
OL: Organizational Learning
OMBOK: Operations Management Body of Knowledge
OSC: Online Synchronous Communications
OSHA: Occupational Safety and Health Administration
PDM: Precedence Diagram Method
PE: Project Evaluation
PEL: Project Evaluation Leadership
PEM: Project Evaluation Management
PG: Process Group
PILE: Personality, Intuition, Language, and Emotions
PM: Project Management
PM&E: Project Management and Evaluation
PMBOK: Project Management Body of Knowledge
PMI: Project Management Institute
PMND: Project Management Network Diagram
PMO: Project Management Office
POLAR: Path of Least Anticipated Resistance
PRINCE2: Project in Controlled Environments (version 2)
PT: Project Team
PTM: Project Team Member
RLC: Records Life Cycle
RM: Records Management
RME: Research, Measurement, and Evaluation
ROI: Return on Investment
SCM: Supply Chain Management
SME: Subject Matter Expert
SNA: System Needs Assessment
SOC: Synchronous Online Communication
SOE: Standard Operating Environment
STACK: Skills, Talent, Aptitude, Capabilities, and Knowledge
SUCCESS: Sustained Understanding Common Critical Expert Strategies
VSM: Value Stream Management
V&P: Visualization and Presentation
WANTS: Wish Lists, Add-ons, Nice to haves: Trinkets and Specialties
WBS: Work Breakdown Structure

Terms and Definitions

(Related concepts not elaborated upon in text)

Acceptance Criteria: Requirements that a project component, i.e., lesson must demonstrate before stakeholders will accept delivery represents a criteria for acceptance.

Acknowledgement: A computer recognizing a request of a user, i.e., electronic signature.

Active Listening: Paying close attention to what is said, asking the other party to describe carefully, clearly, and concisely what is meant in the lesson, and, if necessary, requesting that ideas be repeated to clarify any ambiguity, uncertainty, or anything else that was not understood. Active listening is best performed in person when the person's nonverbal communication (body language) also can be interpreted.

Analysis Paralysis: This term refers to the human inability to effectively make decisions. The individual isn't making the desired progress because of being bogged down in details, i.e., making changes or adjustments. A common approach to addressing this state is to identify, evaluate, design solutions, create a test case, and attempt to remove the causes of the problem. Sometimes this is not possible because the causes cannot be agreed upon or found.

Assignable Cause: A source of variation in a process that can be isolated, especially if it has a significantly larger magnitude or different origin, which distinguishes it from random causes of variation.

Balanced Scorecard: A model method of evaluating business performance based on measures of financial performance, internal operations, innovation and learning, and customer satisfaction.

Band-Aid: A temporary fix or solution to a problem, e.g., loss in network connectivity.

Baseline: A baseline is a "snapshot" in time of one version of each document in a project repository. It provides an official starting point on which subsequent work is to be based, and to which only authorized changes can be made. After an initial baseline is set, every subsequent change to a baseline is recorded as a delta until the next baseline is identified.

Blank Check: Authorization to spend whatever it takes within reason to fix a problem.

Boilerplate: A template or predesigned format that should be followed.

Breach: Unauthorized, unlawful, or illegal practices associated with computer access or use is a breach. To reduce breaches, organizations have disclaimers that prevent inappropriate use.

Business Case: Justification of why the project is necessary and what the deliverables are going to be. It should minimally include key items, such as goals and objectives, and address issues such as Return on Investment (ROI), project risks, and alternatives. The project's sponsor in conjunction with the project team is responsible for developing the business case.

Cascading: The process of developing integrated Scorecards throughout an organization. Each level of the organization develops a scorecard based on the objectives. For example, an IT department could be based on system uptime. Cascading allows every employee and/or department to demonstrate a contribution to overall organizational objectives.

Change Control: Process of redefining time frames, costs, resources, or deliverables associated with a project usually in response to scope change or risk assessment.

Change Management: Systematic approach to implementing agreed upon steps (processes and procedures) to ensure that changes are implemented in an anticipated, planned, and orderly fashion. It may involve instilling new values, encouraging new attitudes, embracing new norms, and creating desired behaviors. It also involves building consensus among employees, and may extend to relationship building with stakeholders and customers. There are various dimensions to change management. One involves sponsorship, which engages sponsors and key stakeholders in the process. Another dimension involves personal resilience, or the ability to adapt to change.

Checklist: A checklist is a tool that is used to determine if an item under investigation meets predefined standards and can be constructed as an outline to enable quick data collection. A checklist also can be created in a matrix design to allow for multidimensional analysis, i.e., ratings per item. It is not uncommon to use special symbols or coding schemes to further compress data collection.

Collaboration: The process of bringing people, processes, or systems together. It involves developing a mutually beneficial, well-defined relationship. Collaboration involves joint planning, sharing resources, and integrated resource management. It also can involve bringing together content from across an organization and between companies. Collaboration tools allow users to seamlessly exchange documents, ideas, and project plans, ideally in real time and from remote locations. Information can be stored in a central repository where it can be shared.

Counter: A software feature that tracks the number of times a Web page is visited is a counter.

Crashing: A project management process that takes action to decrease the total project duration after analyzing a number of alternatives to determine how to get the maximum duration compression for the least cost.

Criteria (criterion): Standards, values, measures, etc., used to help make a determination.

Critique: Critiquing something involves a constructive criticism of the effectiveness or efficiency of a policy, procedure, process, or program.

Cursory Review: A cursory review is a "fake evaluation." It is casual, hasty, incomplete, incompetent, and without attention to detail. It is not planned and not thorough.

Cycle: An established schedule or time period that is recurring for a particular event, e.g., preventative equipment maintenance that is performed annually.

Data Migration: Moving data from one platform (i.e., operating system) to another.

Data Mining: The function of extracting useful or desired information from large data sources.

Data Purging: Removal of data from a rewriteable storage media, e.g., hard drive, in such a way that there is a significant degree of assurance that the data may not be rebuilt.

Data Warehouse: A repository of an organization's historical data and corporate memory. It contains the data necessary to support a decision support system. A major benefit of a data warehouse is that a data analyst can perform complex queries and analysis, such as data mining.

Diagram: A form of visualizing data or information for the purpose of increasing understanding, e.g., decision trees, fishbone diagrams, bar graphs, pie charts, or scatter plots.

Diary: A time-based journal used by the project team to record information, i.e., lessons.

Digital Asset Management: Collecting, cataloguing, storing, and distributing digital content.

Digital Content Delivery: Allowing access to or providing digital files electronically.

Digital Rights Management: Technologies used by publishers (copyright holders) to control access/usage of digital data, and restrictions associated with specific instances of digital works.

Early Adopter: An individual or group who embraces technology from its introduction, and who may be willing to accept, for example, computer system problems or issues in exchange for the perceived benefits of technology and innovation.

Editorial: A statement of opinion or fact by an evaluator regarding a particular issue is an editorial. Editorials are sometimes used to respond to comments made by users on Web sites.

Enterprise Content Management: Systems that enable capture, storage, security, revision control, retrieval, distribution, and preservation of documents.

e-Room: There are a variety of electronic rooms that provide a virtual place where people can communicate in real time while on the Internet or on an intranet. Depending on the configuration, messages can be sent, video can be enabled, and file sharing can occur.

Experiential Learning: Education and learning that is primarily obtained through lived experience.

Expert Systems: These systems are a part of the artificial intelligence category. The computer is able to make decisions based on decisions it has made in the past. This model is powerful in applications, such as managing financial situations, such as when to buy and sell shares on the stock market. The data stored in an expert system is called the knowledgebase. The part of the system that performs decision-making tasks is called the inference engine.

Fault Tolerant: The capability to continue operation after hardware or software failure and without interruption indicates that the system is fault tolerant.

Feasibility: Extent to which a study or project may be done effectively and/or efficiently.

File Compression: Compressing files reduces the amount of storage data required; specifically relates to the design of the data that enables compression. Methods include replacing blank spaces with a character count, or replacing redundant data with shorter stand-in "codes." No matter how data are compressed, they must be decompressed before they can be used.

File Extension: Last few letters, preceded by a period, which identifies a data file's format.

File Structure: Organization of files in directories and subdirectories.

Filtering: A process that executes a function in response to a user request to deliver only relevant predefined information. For example, selecting income level as a filter in a database query will limit the display of records to a specified income level.

Fishbone Diagram: Also called Ishakawa diagram, is a technique used to organize elements of a problem or situation to aid in the determination of causes of the problem or situation.

Framework: A logical structure for classifying and organizing complex information.

Game Plan: Refers to an overall approach (project plan) to achieve the project objective.

Garbage In Garbage Out (GIGO): The concept behind GIGO is that if the input data are wrong, inaccurate, or lack substance, the output data will be the same.

Gateway: A system that transfers data between applications or networks. A gateway reformats the data so it is compatible for the receiving network.

For example, gateways between e-mail systems allow users to exchange messages.

Generalizability: The extent to which lessons from a project collected in one setting can be used to reach valid determinations about how it will perform in other settings.

Graphic Resolution: Refers to the level of display or print quality of graphics. Factors, such as file size, number of pixels, etc., are determining characteristics for resolution quality.

Graphical User Interface (GUI): Interfaces, such as Microsoft Windows, that incorporate colorful windows, dialog boxes, icons, pull-down menus, etc.

Guarantee: An enforceable warranty on the stated capabilities of a product provided certain circumstances are met. A manufacturer's guarantee will either replace or refund all or part of the costs depending on the type of guarantee. For this reason, a "30-day satisfaction guarantee" allowing for return if dissatisfied is desirable, but does not replace the need for a warranty.

Handshake: An established connection between two computer systems.

Harmonization: Process of increasing the compatibility and functionality of something across an organization with the goal of improving performance while supporting conformity (similar). It is somewhat different than standardization, which is concerned with uniformity (same).

Historical Record: Documentation of previous activities or events, e.g., lessons.

Honor System: An agreement between users that they can be trusted to act in a dependable manner regarding behavior. Honor systems are usually unsupervised and operate under the belief that people will not take unfair advantage of others, e.g., use of computer equipment.

Human capital management: Realizing the investment an organization makes in its employees and focusing on productivity per employee.

Human Resources Information System (HRIS): Is adaptable to embody full-scale human capital management, i.e., employee development planning, job performance evaluation, compensation, coaching, and talent management.

Hybrid Approach: Using customized or tailored methods to accomplish a task.

Hypothesis: A possible explanation of an observation or an event.

Information Mapping: Converting large amounts of text to smaller amounts of text using tables and formatting.

Informed Consent: Providing clients with sufficient information of a study who are to participate in so they understand the risks and benefits of the issues.

Instance: An occurrence or happening of an event. The more instances that are reported, the more a pattern can be predicted.

Instrument: A tool such as a checklist or survey that is used to complete project-related work.

Intellectual Property: The rights of the author under copyright law designed to protect proprietary knowledge, such as any published works of literature, in whatever form.

Intelligent Information Management: A component of information infrastructure that helps manage digital information. It can be set up to automatically discover information and assess its importance. It reportedly lowers costs, reduces risks, and creates new uses for information.

Interaction: An event that involves some physical connection between two people.

Isomorphic Learning: Universally applicable lessons that have obtained an analysis of factors that can be used to address similar future situations.

Jeopardy: A condition that represents major potential concerns on a project. It may result in the project team developing alternative solutions to mitigate the risk.

Job description: Documentation of a person's job title, roles, responsibilities, authority level, and related job duties. It outlines desired education, work experience, and competencies. It also may include projected career path, reporting structure, and related information.

Joint Committee on Standards (JCOS): Support for the development of personnel, program, and student evaluation standards. JCOS was founded in 1975 by the American Educational Research Association, the American Psychological Association, and the National Council on Measurement in Education. JCOS now includes many other organizations in its membership. AEA is one of those organizations and has a representative to the joint committee.

Journal: A record (handwritten or typed) that is completed on a periodic basis to document experiences, observations, activities, or events.

Juggle: To handle multiple competing priorities or project tasks simultaneously.

Just-in-Case: Contingency planning solutions that offer alternatives.

Just-in-Time: Instruction available when an individual needs it in order to perform an activity rather than at the time someone else wants to make it available.

Keyed-in: The data have been entered into the computer.

Keyword: A term or phrase used as an identifier. Search engines use keywords to locate information on the Internet. Keywords are the essential ingredient to any effective content management system.

Kickback: Anything of value, e.g., money, or additional future work where coercion is involved between individuals with the intent to change the outcome of a project evaluation.

Kickoff Meeting: The first meeting with the project team and project stakeholders.

Knowledge Base: A base of knowledge or the sum of knowledge that has been accumulated and is available for dissemination to the organization through their data management systems.

Knowledge Mapping: A process of categorizing knowledge as information or association; usually put into a matrix or database as part of the mapping process.

Knowledge Marketplace: Internet sites where subscribers can share, purchase, buy, or exchange knowledge on specific material.

Knowledge Objects: Tools, utilities, and resources that are used to access an organization's knowledge management system.

Launch: A starting point for project initiatives

Lean: A streamlined method of performing tasks

Learning and Development: Also referred to as training, is responsible for processes, programs, policies, and procedures used to educate, inform, and instruct people.

Legend: A key or reference that is used to define a specific attribute of an item. The key may contain symbols, shapes or colors that reference an item on a graph. For example, red may be used to indicate risk; whereas green may indicate an acceptable state.

Level of Effort: The amount of energy (physical and mental), communication, time, resources (people, equipment, facilities, and supplies) required to complete a task.

License: A legal document that allows persons or organizations to use hard or software.

Life Cycle: The typical process a project undergoes from start to completion. For example, in a project, it is initiated, then planned and thereafter executed. Finally, it is closed. This represents a four-stage life cycle.

Logic: A way of processing human thought in a sequential manner or step-by-step.

Logistics: The turnkey process of starting a project and completing it, taking into consideration all required resources, i.e., people, equipment, facilities, systems and materials. The use of this term is context specific. For example, in supply chain management, logistics involves managing the flow of materials and services between the point of origin and the point of use in order to meet customer requirements.

Management by Walking Around (MBWA): A management technique that involves physical observation and oversight to ensure requests are being followed.

Maturity Model: Used to assess an organization's progress, i.e., level of knowledge. It is a benchmarking instrument that can help an organization grow. It is common to see maturity models in functional areas, such as project management or supply chain management.

Meeting Minutes: The written record of a meeting or related communication that provides details on discussion topics and action items.

Metrics: Items that are measured to better understand performance levels. Metrics are quantitative, but can include qualitative factors. Metrics increase in value as they are taken over time and monitored to assess performance against criteria, e.g., sales increase.

Middleware: Software that translates information between two applications.

Milestone: A significant event in the project, usually completion of a major deliverable.

Mirror: A backup copy of a directory on a media, e.g., a hard disk drive.

Misevaluation: When evaluators perform poorly, inadequately, unethically, untruthfully, and/or fail to meet evaluation objectives results in a misevaluation.

Naïve: A state of being unaware or inexperienced due to a lack of information or interest.

Narrative: A format of creating lessons in which the information is descriptive and composed in sentence format.

Native Format: The original format the data (document) was originally created in is considered native. For example, the document was created in MS Word and was saved as an MS Word .doc or other compatible format inherent as an output/save option in MS Word.

Node: A point on the network, such as a personal computer, server, mainframe, and, in some cases, peripheral are considered nodes.

Nonsense: A classification of information that does not have perceived relevance or validity.

Notation: Adding a comment to an existing document.

Notification: Notifying is the process of communicating the status of a system action. For example, autonotification may be set to "on" each time the system receives a request for a report to let the system administrator know what report was requested.

Nothing: The absence of anything, being totally void where something can be extracted. In the context of data or information, it may represent insignificance, while it, in fact, may exist.

Object Linking and Embedding (OLE): A method of linking information between software applications. When the spreadsheet is updated in MS Excel, then it is automatically updated in the MS Word document.

Obsolescence: The loss of value, usefulness, functionality, compatibility resulting from advances in technology and the passage of time.

Offline: Something that is not presently active or available for access in a system; in some contexts also may refer to a system that is down for maintenance or inoperable.

Online Analytical Processing (OLAP): Capability for databases and related processes to handle more complex queries than traditional relational databases.

Online: Something that is active or available for access in a system; also may refer to connected to a network versus being stand-alone.

Organizational Readiness: The ability for an organization to embrace change and continuously adapt to technological advances that are integral to its business functions. For example, an organization's ability to accept changes in computer operating systems and software as the industry migrates to new platforms to keep up with the times.

Over-the-Top: When something is unacceptable or classified harshly, i.e., ridiculous.

Performance Needs Assessment (PNA): A method to determine the relative conformance of a person or system to prescribed criteria. There are the four Ds to conducting a PNA: (1) Define desired performance. Ask, "What is the worker (or system) expected to do?" "How well is the worker (or system) expected to perform," "Under what circumstances," "How often." (2) Describe actual performance. The difference between desired or anticipated performance and actual performance is called the performance gap. (3) Design and conduct a root cause analysis to find out why there is a performance gap. Gather information from individuals (or systems). And, (4) Determine appropriate intervention(s) to improve performance. If the cause is insufficient knowledge (or a slow system), enhancing training (or upgrading to a faster processor) may be the appropriate intervention.

Phenomenon: A unique occurrence or event that represents significance over the usual. In the context of a lesson, it should be captured due to its uniqueness.

Plus R/Minus R Positive Reinforcement (+R/-R): +R/-R is an internationally recognized approach to improving performance through positive or negative reinforcement.

Pollution: The socio-political elements that arise during a project that cloud results.

Profile: The characteristics or attributes of an item under study.

Protocol: Rules, standards, and guidelines that govern how an activity will be completed.

Qualitative Analysis: Based on observation, professional judgment, experience, investigation, questioning, or similar methods. It is an analysis that attempts to determine the nature of things, attributes, values, behaviors, impacts being measured. It seeks to find the things that are tangible and intangible. Qualitative research is said to be exploratory and look at why and how. Qualitative reports are generally narrative in form, but can use some graphics for presentation.

Quality Assurance (QA): Systematic monitoring, control, and evaluation of a project.

Quantitative Analysis: Using mathematical and/or numerical representation of data for the purpose of describing the item under study. For example, in a training scenario, analyzing training participants' scores on a test can be a quantitative approach. Quantitative approaches use numbers, amounts, values, percentages, ratios, portions, etc., in the description of the phenomena. Because of the need to present measurable results, statistics is employed.

Query: A manual entry by a user who enters a character, word, phrase, or wildcard representing the information a user seeks from search engines and directories. The search engine subsequently attempts to locate the requested information within directories or files.

Quorum: The minimum number of participants (commonly voting board members) that must be present before a decision can be made.

Quote: An estimate for a project that comes from a vendor involved in solicitation.

Read Receipt: Process of requesting verification that an e-mail has been seen is a read receipt.

Recovery: Attempt to restore parts or all of a system's information due to a failure or error.

Recycle Bin: The final storage location of files that have been deleted, which can be recovered.

Redundancy: Duplication or back-up that provides support for business continuity.

Refresh Rate: The length of time required for new information to appear on screen.

Reinventing the Wheel: Performing the same work tasks over again instead of smart reuse.

Remediation: An action taken to correct a problem or fix a situation.

Requirements Document: A formal planning document approved by the sponsor and project manager, it establishes the framework for the project, clearly outlining tasks and expectations.

Resilient: A person's ability to adapt to change.

Responsibility Assignment Matrix (RAM): A document that specifies roles and responsibilities with specific dates and objectives for completion is an accountability matrix.

Side Effect: A secondary impact (usually negative) that may be unanticipated.

Sponsor: Person who has oversight for the project, e.g., owner or financier.

Statement of Work (SOW): A narrative description of services supplied under agreement.

Storytelling: A narrative form of describing lessons learned by giving a related example to make the situation more descriptive and relevant. Storytelling may be fact or fiction.

Structured Query Language (SQL): Industry standard language used by software to access a database. There is a common instruction set that works with most databases.

Subevaluation: A subset or part of the main evaluation.

Subscriber: A registered user, typically provided with a user name and password.

Synthesis: Combination or integration of elements to form a whole.

System Integration Testing: Testing hardware or software for compatibility and performance.

System Specification: Outlines the system description, functional requirements, technical parameters, design constraints, and acceptance criteria.

Testing: An examination of the attributes or characteristics of a person, process, or system, usually in a sequential manner. There are different types and levels of tests administered, depending upon requirements. For example, testing of people is performed in a written, oral, or hands-on method to determine if a desired level of skill, knowledge, or ability is present. Process testing is concerned with validity and reliability. System testing is designed to address performance, e.g., stress tests, which attempt to determine dependability by analyzing factors, such as Mean Time Between Failure (MTBF) and Mean Time To Repair (MTTR).

Theory: A well-founded explanation that has undergone comprehensive review.

Total Quality Management (TQM): A management philosophy that attempts to implement a quality improvement program, process, or procedure within an organization.

Transparency: Open, honest, and direct communication throughout the organization.

Treatment: Solutions that serve to address a problem condition in part or whole.

Trial: A user-based evaluation to determine if an item, e.g., product, meets desired specifications.

Triple Constraint: Factors that must be adhered to in a project include time, cost, and scope.

Understanding: The mental state of being where ways of knowing an issue is concrete.

Uniformity: The point at which processes are consistent without noticeable variation.

Uptime: The amount of time that the system is available to users in a fully functional state.

User Acceptance Testing (UAT): Testing that is designed to ensure usability of a system or process. For example, Alpha testing refers to the first phase and is usually conducted by the development team. Beta testing refers to the second phase and is usually conducted by subject matter experts before deployment to the target audience.

User Friendly: Usually refers to software applications, and is considered intuitive, easy to use, visually attractive, meeting or exceeding end-user expectations; processing is considered quick; output is in a format that is acceptable; data is accurately stored and easy to retrieve. Features can be customized to the user's liking and adaptable to the work environment.

Utility: Usefulness of the data or information to its intended users.

Utilization: The amount, degree, or extent to which an item is consumed or used.

Utopia: An imaginary state of being accompanied by joy that lasts for a short period of time, and which is frequently experienced with the installation of new hardware and software.

Validation: Giving something the stamp of approval or rejection; concerns adhering to specifications under the context of specific parameters. Validation involves checking data entry for accuracy. A check is made to ensure that data meet specified criteria, i.e., being of the correct category. Validation allows data to be checked before processing takes place.

Verification: Involves ensuring that the information transferred, or copied from a source, is the same as the original source documents. Verification techniques used at the basic level include looking at the date and time a document was created and document size. Advanced verification includes looking at every character within a document in addition to file size and date created.

Version: Also referred to as release, update, revision, or build. Version control is the management of multiple changes or updates of the same unit of information. Version control tools are particularly important in publishing, saving, retrieving, and editing content. Keeping system users current with the same version is critical to data management.

Virtual Collaborative Environment (VCE): To work together from a remote location using technologies, such as Web conferencing, video conferencing, and teleconferencing.

Virtual Organization: A short-term, periodic, or temporary alliance between independent organizations in a potentially long-term relationship. Organizations cooperate based on mutual values and act as a single entity to third parties.

Warranty: A promise that the product, e.g., a computer, will meet a specified level of performance over a specified period of time; may apply to a product that is older or damaged in some respect, but will function to the desired level.

Watermark: A way of putting a signature on an image indicating its authorship.

Webinar: A Web-based seminar usually intended for orientation, training, presentation, or meetings; usually incorporates a teleconference and PowerPoint-type presentation.

White Paper: An authoritative report on a major issue whose purpose is to educate or inform. Lessons are commonly reformatted as white papers.

Wiki: A software program that allows users to create, edit, and link Web pages. Collaborative Web sites and community Web sites are often created with wikis.

Wildcard: A special character, such as an asterisk (*), that you can use to represent one or more characters.

Windows, Icons, Menus, and Pointers (WIMP): Graphical user interfaces, such as Windows.

Wizard: A software function that attempts to determine a sequence of desired steps and then execute those steps in an interactive fashion with the user.

X-rated: Classifying content within a lesson as restricted, obscene, or offensive. Lessons that fit this category may include inappropriate employee behavior. Lessons with this classification should be identified and restricted to specific audiences, e.g., human resources and legal.

X-ray: Refers to a deep review of circumstances or conditions to determine underlying causes. Multiple systems, processes, or resources may be involved to determine diagnosis. Evaluations that involve financial considerations where a breach in trust has occurred may fit this profile.

Yes-Man: An individual, usually in a semiinfluential position, that agrees to circumstances or decisions verbally, even when, in actuality, they do not, for the sole purpose of organizational politics.

Yo-yo: A state (unlike analysis paralysis) where decisions regarding lessons go up and down, where no solid determination can be made, and when things make no sense.

Zealous: The ideal state for project team members to be in as they engage in postproject reviews. It is an emotional state that extends beyond engaged to a purposeful sense of commitment and enjoyment when appropriate.

Zero-Defect: A quality standard that reinforces a no-fault philosophy.

Evaluation

Alkin, M. 2004). Evaluation Roots: Tracing Theorists Views and Influences. Thousand Oaks, CA: Sage Publications.

Altschuld, J. 1999). The certification of evaluators: Highlights from a report. American Journal of Evaluation, 20, 481–493.

Altschuld, J., & Wikin, B. 2000). From Needs Assessment to Action: Transforming Needs Into Solution Strategies. Thousand Oaks, A: Sage Publications.

American Evaluation Association. 2007). Guiding principles for evaluators.

Bamberger, M., Rugh, J., & Mabry, L. 2006). *Real world evaluation: Working under budget, time, data, and political constraints.* Thousand Oaks, CA: Sage Publications

Bichelmeyer, B. 2003). Checklist for formatting checklists.

Brinkerhoff, R. 2006). Telling training's story: Evaluation made simple, credible and effective—using the success case method to improve learning and performance. San Francisco, CA: Berrett-Koehler Publishers, Inc.

Butler, J. 2005). Metaevaluation and implications for program improvement. Retrieved July 06, 2007 from http://www.acetinc.com/Newsletters/Issue%2010.pdf.

Carmines, E. & Zeller, R. 1979). Reliability and Validity Assessment. Thousand Oaks, CA: Sage Publications.

Cook, T., Levinson-Rose, J., & Pollard, W. 1980). The misutilization of evaluation research: some pitfalls of definition. Science Communication, 1, 477–498.

Davidson, J. 2005). Evaluation methodology basics: The nuts and bolts of sound evaluation. Thousand Oaks, CA: Sage Publications

Donaldson, S. & Scriven, M. 2003). Evaluating Social Programs and Problems: Visions for the New Millennium. Mahwah, NJ: Lawrence Erlbaum & Associates.

Edwards, J., Scott, & J., Raju 2003). The Human Resources Program-Evaluation Handbook. Thousand Oaks, CA: Sage Publications.

Fetterman, D., Kaftarian, & S. Wandersman 1996). Empowerment Evaluation: Knowledge and ools for Sel-Assessment & Accountability. Thousand Oaks, CA: Sage Publications.

Fitzpatrick, J. Sanders, J. & Worthen, B. 2004). Program Evaluation 3rd Edition: Alternative Approaches and Practical Guidelines. Pearson: New York, NY.

Forss, K. Cracknell, B., & Samset, K. 1994). Can evaluation help an organization to learn? Evaluation Review, 18, 574–591.

Frechtling, J. 2007). Logic Modeling Methods in Program Evaluation. San Francisco: CA: Jossey-Bass.

Gajda, R. 2004). Utilizing collaboration theory to evaluate strategic alliances. American Journal of Evaluation, 25, 65–77.

Hale, J. 2002). Performance Based Evaluation: Tools and Techniques to Measure the Impact of Training. San Francisco, CA. Josey-Bass Pfeiffer.

Hunt, M. 1997). How Science Takes Stock: The Story of Meta-Analysis. New York, NY: Russell Sage Foundation.

Joint Committee on Standards for Educational Evaluation. 1988). The student evaluation standards: How to assess evaluations of educational programs. Thousand Oaks, CA: Sage Publications

Joint Committee on Standards for Educational Evaluation. 2004). The program evaluation standards: How to assess evaluations of educational programs, 2nd ed. Thousand Oaks, CA: Sage Publications

Joint Committee on Standards for Educational Evaluation. 1994. The program evaluation standards: How to assess evaluations of educational programs, 2nd ed. Thousand Oaks, CA: Sage Publications

Kane, M. & Trochim, W. 2007 Concept Mapping for Planning and Evaluation. Thousand Oaks, CA: Sage Publications.

Kirkpatrick D. and Kirkpatrick J. 2006. Evaluating Training Programs. San Francisco, CA: Berret-Koehler Publishers.

Knowlton, L. & Phillips C. 2009. The Logic Model Guidebook. Thousand Oaks, CA: Sage Publications

Kusek, J. & Rist, R. 2004. Ten steps to a results-based monitoring and evaluation system. Washington, DC: World Bank Publications.

Lawrenz, F., & Huffman, D. 2003. How can multi-site evaluations be participatory? American Journal of Evaluation, 24, 471–482.

Lipsey, M. & Wilson, D. 2001. Practical Meta-Analysis Thousand Oaks, CA: Sage Publications.

Littell, J., Corcoran, & J., Pillai, V. 2008. Systematic Reviews and Meta-Analysis. New York, NY: Oxford University Press.

Mathison, S. Ed. 2005. Encyclopedia of evaluation. Thousand Oaks, CA: Sage Publications

McDavid, J. & Hawthorn, L. 2006. Program Evaluation * Performance Measurment: An Introduction to Practice. Thousand Oaks, CA: Sage Publications.

Patton, M. 1997. Utilization-Focused Evaluation 3rd Edition: The New Century Text. Thousand Oaks, CA: Sage Publications.

Preskill, H. & Russ-Eft 2005. Building Evaluation Capacity: 72 Activities for Teaching and Training. Thousand Oaks, CA: Sage Publications.

Preskill, H., & Catsambas, T. 2006. *Reframing evaluation through appreciative inquiry.* Thousand Oaks, CA: Sage Publications.

Preskill, H., & Torres, R. 1999. *Evaluative inquiry for learning in organizations.* Thousand Oaks, CA: Sage Publications.

Raupp, M., & Kolb, F. 1990. Evaluation management handbook. Andover, MA: Network, Inc.

Reineke, R., & Welch, W. 1986. Client centered metaevaluation. American Journal of Evaluation, 7, 16–24.

Reeve, J., & Peerbhoy, D. 2007. Evaluating the evaluation: Understanding the utility and limitations of evaluation as a tool for organizational learning. Health Education Journal, 66, 120–131.

Renger, R., & Titcomb, A. 2002. A three-step approach to teaching logic models. American Journal of Evaluation, 23, 493–503.

Rist, R. & Stame, N. 2006. From Studies to Streams: Managing evaluative systems. New Brunswick, NJ: Transaction Publishers.

Rodriguez-Campos, L. 2005. Collaborative evaluations: A step-by-step model for the evaluator. Tamarac, FL: Llumina Press.

Rosas, S. 2006. A methodological perspective on evaluator ethics. American Journal of Evaluation, 27, 98–103.

Rossi, P., Lipsey, M. & Freeman, H. 2004. Evaluation A Systematic Approach 7th Ed. Thousand Oaks, CA: Sage Publications.

Royse, D., Thyer, B, Padget, D. & Logan T. 2006. Program Evaluation 4th Ed: An Introduction. Belmont, CA: Thomson Higher Education.

Ruhe, V. & Zumgo, B. 2009. Evaluation in Distance Education and E-learning. New York, NY: The Guilford Press.

Schwandt, T. & Halpern, E. 1990. Linking auditing and metaevaluation: Enhancing quality. Applied Research, 10, 237–241.

Scriven, M. 1991. Evaluation thesaurus, 4th ed. Newbury Park, CA: Sage Publications.

Shaw, I., Greene, J., Mark, M., 2006. The Sage Handbook of Evaluation. Thousand Oaks, CA: Sage Publications.

Stevahn, L., King, J., Ghere, G., & Minnema, J. 2005. Establishing essential competencies for program evaluators. American Journal of Evaluation, 26, 43–59.

Straw, J. 2002. The 4-Dimensional Manager: DISC Strategies for Managing Different People in the Best Ways. San Francisco, CA: Berrett-Koehler Publishers, Inc.

Stufflebeam, D. 2000. Guidelines for developing evaluation checklists. Retrieved September 10, 2007 from http://www.wmich.edu/evalctr./checklists /checklists-menu.htm.

Stufflebeam, D. 2001a. Evaluation checklists: Practical tools for guiding and judging evaluations. American Journal of Evaluation, 22, 71–79.

Stufflebeam, D. 2001b. The metaevaluation imperative. American Journal of Evaluation, 22, 183–209.

Stufflebeam, D. 2001. Evaluation Models. San Francisco, CA: Jossey-Bass.

Torres, R., Preskill, H. & Piontek, M. 2005. Evaluation Strategies for Communicating and Reporting 2nd Ed. Thousand Oaks, CA: Sage Publications.

Witkin, B. & Altschuld, J. 1995. Planning and Conducting Needs Assessment. Thousand Oaks, CA: Sage Publications.

Wolf, F. 1986. Meta-Analysis – Quantitative Methods for Research Synthesis. Thousand Oaks, CA: Sage Publications.

Yang, H., & Shen, J. 2006. When is an external evaluator no longer external? Reflections on Some Ethical Issues. American Journal of Evaluation, 27, 378–382.

Knowledge Management

Borghoff, U., & Pareschi, R. 1998. Information technology for knowledge management. Germany: Sprinter-Verlag.

Brockmann, E. & Anthony, W. 2002. Tacit knowledge and strategic decision making. Group Organization Management, 27, 436–455.

Chua, A., Lam, W., & Majid, S. 2006. Knowledge reuse in action: The case of call. Journal of Information Science, 32, 251–260.

Chou, S., & He, M. 2004. Knowledge management: The distinctive roles of knowledge assets in facilitating knowledge. Journal of Information Science, 30, 146–164.

Chou, T., Chang, P, Tsai, C., & Cheng, Y. 2005. Internal learning climate, knowledge management process and perceived knowledge management satisfaction. Journal of Information Science, 31, 283–296.

Hall, H. 2001. Input-friendliness: Motivating knowledge sharing across intranets. Journal of Information Science, 27, 139–146.

Kaiser, S., Mueller-Seitz, G., Lopes, M., & Cunha, M. 2007. Weblog-technology as a trigger to elicit passion for knowledge. Organization, 12, 391–412.

Kalseth, K., & Cummings, S. 2001. Knowledge management: Development strategy or business strategy. Information Development, 17, 163–172.

Knowles, M., Holton, E. & Swanson, R. 1998. The Adult Learner. Houston, TX: Gulf Publishing.

Marquardt, M. 1996. Building the Learning Organization. New York, NY: McGraw Hill.

Nelson, S. 1979. Knowledge creation: An overview. Science Communication, 1, 123–149.

Reich, B. 2007. Managing knowledge and learning in IT projects: A conceptual framework and guidelines for practice. Project Management Journal, June, 5–17.

Tiwana, A. 2002: The knowledge management toolkit: Orchestrating IT strategy and knowledge platforms. Upper Saddle Ridge, NJ: Pearson Publications.

Torraco, R. 2000. A theory of knowledge management. Advances in Developing Human Resources, 2, 38–62.

Walsham, G. 2002. What can knowledge management systems deliver. Management Communication Quarterly, 16, 267–273.

Yang, C., & Chen. L. 2007. Can organizational knowledge capabilities affect knowledge sharing behavior? Journal of Information Science, 33, 95–109.

Yates-Mercer, P., & Bawden, D. 2002. Managing the paradox: The valuation of knowledge and knowledge management. Journal of Information Science, 28, 19–29.

Lessons Learned

Abramovici, A. 1999. Gathering and using lessons learned. PM Network, October, 61–63. Philadelphia, PA: Project Management Institute.

Berke, M. 2001. Best practices lessons learned BPLL: A view from the trenches. Proceedings from the Project Management Institute. Philadelphia, PA: Project Management Institute.

Bucero, A. 2005. Project know-how. PM Network. May, 22–23. Philadelphia, PA: Project Management Institute.

Cowles, T. 2004.Criteria for lessons learned: A presentation for the 4th annual MMI technology conference and user group, November 16, 2004. Denver, CO: Raytheon, Inc. Retrieved on August 11, 2011 from www.dtic.mil/ndia/2004cmmi/CMMIT2Tue/LessonsLearnedtc3.pdf.

Darling, M., Parry, C., & Moore, J. 2005. Harvard Business Review. Learning in the Thick of It. Boston, MA: Harvard Business School Press.

Darling, M., Parry, C. Growing Knowledge Together: Using Emergeny Learning and EL Maps for Better Results. The SOL Journal. Vol 8 No. 1.

Estrella, J. 2001 . Lessons Learned in Project Management: 140 Tips in 140 Words or Less. Charleston, SC: CreateSpace.

Friedman, V., Lipshitz, R., & Popper, M. 2005. The mystification of organizational learning. Journal of Management Inquiry, 14, 19–30.

Goodrum, P., Yasin, M., & Hancher, D. 2003. Lessons Learned System for Kentucky Transportation projects. Lexington, KY: University of Kentucky.

Grabher, G. 2004. Learning in projects, remembering in networks: Communality, sociality, and connectivity in project ecologies. European, Urban and Regional Studies, 11, 103–123.

Hall, R. 2009. Master of Disaster. PM Network. Philadelphia, PA: Project Management Institute.

Harrison, W., Heuston, G., Morrissey, M. Aucsmith, D. Mocas, S., & Russelle, S. 2002. A lessons learned repository for computer forensics.

Hatch, S. 2011. Lessons Learned From Going Hybrid: Three Case Studies. Retrieved February 27, 2011 from http://meetingsnet.com/associationmeetings/news/hybrid_meeting_case_studies_virtual_edge_0209/.

Hildebrand, C. 2006. On-demand education. PM Network. August, 86. Philadelphia, PA: Project Management Institute.

Hynnek, M. 2002. A real life approach to lessons learned. Project Management Innovations, Vol 2, pp. 5–6. Retrieved on August 11, 2011 from dewi.lunarservers.com/~ciri03/pminpdsig/.../November2002NPDSIG.pdf.

Kaner, C, Bach, J. and Pettichord, B. 2001. Lessons Learned in Software Testing. Hoboken, NJ: Wiley Publishing.

Kendrick, T. 2004. The Project Management Tool Kit: 100 Tips and Techniques for Getting the Job Done Right. Saranac Lake, NY: Amacom.

Kozak-Holland, M. 2002. On-Line, on-Time, on-Budget: Titanic Lessons for the E-Business Executive Lessons From History Series. MC Press.

Kotnour, T. 2000. Leadership mechanisms for enabling learning within project teams. Retrieved December 12, 2007 from www.alba.edu.gr/OKLC2002/Proceedings/pdf_files/ID340.pdf.

Ladika, S. 2008. By focusing on lessons learned; project managers can avoid repeating the same old mistakes. PM Network, February.

Leake, D., Bauer, T., Maguitman, A., & Wilson, D. 2000. Capture, storage and reuse of lessons about information resources: Supporting task-based information search. Retrieved on August 11, 2011 from ftp://ftp.cs.indiana.edu/pub/leake/p-00-02.pdf.

Lewkowicz, J. & Nunan, D. 1999. The Limits of Collaborative Evaluation. China: University of Hong Kong.

Lientz, B. & Rea, K. 1998 Project Management for the 21st Century, 2nd Ed. Burlinton, MA: Academic Press

Loo, R. 2002. Journalilng: A Learning Tool for Project Managment Training and Team-building. Project Managment Journal. Vol 33, No. 4 61–66.

MacMaster, G. 2000. Can we learn from project histories? PM Network, July. Philadelphia, PA: Project Management Institute.

Marqquardt, M. 1996. Building the Learning Organization: a Systems Approach to Quantum Improvement and Global Success. New York: Mcgraw Hill.

Newell, S., Bresnen, M., Edelman, L., Scarbrough, H., & Swan, J. 2006. Sharing knowledge across projects: Limits to ICT-LED project review practices. Management Learning, 37, 167.

Oberhettinger, D. 2005. Workshop on NPR 7120.6, the NASA lessons learned process: Establishing an effective NASA center process for lessons learned. Houston, TX: NASA.

Patton, M. 2001. Evaluation, knowledge management, best practices, and high quality lessons learned. American Journal of Evaluation, 22, 329–336.

Perrot, P. 2001. Implementing Inspections at AirTouch Celluslar: Verizon Wireless. Retrieved on August 11, 2011 from http://sasqag.org/pastmeetings/Implementing %20Inspections%20at%20AirTouch%20Cellular%202-2001.ppt

Pitagorsky, G. 2000. Lessons learned through process thinking and review. PM Network. March, 35–38. Philadelphia, PA: Project Management Institute.

Rowe, S., & Sikes, S. 2006. Lessons learned: Sharing the Knowledge. PMI Global Congress Proceedings. Seattle Washington. Philadelphia, PA: Project Management Institute.

Rowe, S., & Sikes, S. 2006. Lessons learned: Taking it to the next level. PMI Global Congress Proceedings. Seattle Washington. Philadelphia, PA: Project Management Institute.

Seningen, S. 2004. Learn the value of lessons learned. The Project Perfect White Paper Collection. Retrieved on August 11, 2011 from www.projectperfect.com.au/downloads/.../info_lessons_learned.pdf

Sharif, M., Zakaria, N., Chign, L., Fung., L. & Malaysia, U. 2005. Facilitating Knowledge Sharing through Lessons Learned System. Journal of Knowledge Management Practice.

Snider, K., Barrett, F., & Tenkasi, R. 2002. Considerations in acquisition lessons-learned system design. Acquisition Review Quarterly, Winter, 67–84.

Speck, M. 1996. "Best Practice in Professional Development for Sustained Educational Change." ERS Spectrum Spring pg. 33–41.

Spilsbury, M., Perch, C., Norgbey, S., Rauniyar, G., & Battaglino, C. 2007. Lessons learned from evaluation: A platform for sharing knowledge.

Stephens, C., Kasher, J. Walsh, A., & Plaskoff, J. 1999. How to transfer innovations, solutions, and lessons learned across product teams: Implementation of a knowledge management system. Philadelphia, PA: Project Management Institute.

Terrell, M. 1999. Implementing a lessons learned process that works. Proceedings of the 30th Annual Project Management Institute Seminars & Symposium. Philadelphia, PA: Project Management Institute.

The Adult Learner: The Definitive Classic in Adult Education and Human Resource Development Gulf Professional Publishing, Trade paperback 1998

The Office of the Whitehouse 2006. The Federal Response to Hurricane Katrina Lessons Learned.

Weber, R., Aha, D., Munoz-Avila, H., & Breslow, L. 2000. An intelligent lessons learned process. Proceedings of the Twelfth International Symposium on Methodologies for Intelligent Systems ISMIS 2000, 358–367.

Weber, R., Aha, D., Becerra-Fernandez. 2000. Categorizing Intelligent Lessons Learned Systems.

Wheatley, M. 2003. In The Know. PM Network. May, 33–36.

Whitten, N. 2007. In hindsight: Post project reviews can help companies see what went wrong and right. PM Network, 21.

Williams, T. 2007. Post-project reviews to gain effective lessons learned. Philadelphia, PA: Project Management Institute.

Williams, T., Eden, C., Ackermann, F., & Howick, S. 2001. The use of project post-mortems. Proceedings of the Project management Institute Annual Seminars and Symposium. November. Philadelphia, PA: Project Management Institute.

50 Lessons. 2007. Lessons Learned—Straight Talk from the World's Top Business Leaders—Leading By Example. Boston, MA: Harvard Business School Press.

Project Management

Avison, D. & Torkzadeh G. 2009. Information Systems Project Management. Thousand Oaks, CA: Sage Publications

Crowe, A. 2006. Alpha project managers: What the top 2% know that everyone else does not. Kennesaw, GA: Velociteach, Inc.

Fabac, J. 2006. Project management for systematic training. Advances in Developing Human Resources, 8, 540–547.

Friedrich, R. 2007. The essence of OPM3.

Heldman, K. 2005. Project management professional study guide, 4th ed. Hoboken, NJ: Wiley Publishing, Inc.

Kerzner, H. 2004. Project Management: A Systems Approach to Planning, Scheduling and Controlling. Hoboken, NJ: John Wiley & Sons.

Kerzner, H. 2009. Advanced Project Management: Best Practices on Implementation. Hoboken, NJ: John Wiley & Sons.

Kerzner, H. & Saladis, F. 2009. Value-Driven Project Management. New York, NY: IIL Publishing.

Llewellyn, R. 2006. PRINCE2 vs. PMP. Retrieved December 21, 2007 from http://manage.wordpress.com/2006/11/24/prince2-vs-pmp/.

Locke, L., Spirduso, W & Silverman, S. . Proposals that Work 5th Edition: A Guide for Planning Dissertations and Grant Proposals. Thousand Oaks, CA: Sage Publications.

Marsh, D. 1996. Project management and PRINCE. Health Informatics, 2, 21–27.

Middleton, C. 1967. How to set up a project organization. Harvard Business Review, March–April, 73–82.

Mulcahy, R. 2005. *PMP exam prep: Rita's course in a book for passing the PMP exam,* 5th ed., Minneapolis, MN: RMC Publications, Inc.

PRINCE2. 2009. Managing Successful Project with Prince2. London, UK: The Stationary Office.

Project Management Institute. 2002. Project manager competency development framework. Newtown Square: PA: Project Management Institute.

Project Management Institute. 2003. Organizational project management maturity model OPM3: Knowledge foundation. Newtown Square: PA: Project Management Institute.

Project Management Institute. 2004. A guide to the project management body of knowledge, 3rd ed. Newtown Square: PA: Project Management Institute.

Project Management Institute. 2006. The Practice Standard for Work Breakdown Sructures. Newtown Square: PA: Project Management Institute.

Project Management Institute. 2007. The Practice Standard for Work Breakdown Sructures. Newtown Square: PA: Project Management Institute.

Project Management Institute. 2002. Project Manager Competency Development Framework. Newtown Square: PA: Project Management Institute.

Project Management Institute. 2008. The Standard for Program Management 2nd Ed. Newtown Square: PA: Project Management Institute.

Project Management Institute. 2008. The Standard for Portfilio Management 2nd Ed. Newtown Square: PA: Project Management Institute.

Project Management Institute. 2007. Source: Project management institute code of ethics and professional conduct.

Project Management Institute. 2006. Government Extension to the PMBOK Guide 3rd Ed. Newtown Square: PA: Project Management Institute.

Project Management Institute. 2004. Professionalization of Project Management. Newtown Square: PA: Project Management Institute.

Rad, P. & Levin, G. 2006. Project Portfolio Management: Tools and Techniques. New York, NY: IIL Publishing.

Sartorius, R. 1991. The logical framework approach to project design and management. American Journal of Evaluation, 12, 139–147.

Stackpole, C. 2009. A Project Manager's Book of Forms. Hoboken, NJ.: John Wiley & Sons.

Stewart, W. 2001. Balanced scorecard for projects. Project Management Journal, March, 38–53.

Zwerman, B., Thomas, J., Haydt, S., & Williams, T. 2004. Professionalization of project management: Exploring the past to map the future. Philadelphia, PA: Project Management Institute.

Records Management

Bates, S., & Smith, T. 2007. SharePoint 2007 user's guide: Learning Microsoft's collaboration and productivity platform. New York, NY: Springer-Verlag.

Board of Regents University of Wisconsin System. 2002. Basics of Good Evaluation Reporting. Madison, WI: University of Wisconsin-Madison.

Cohen, D., Leviton, L., Isaacson, N., Tallia, A., & Crabtree, B. 2006. Online diaries for qualitative evaluation: Gaining real-time insights. American Journal of Evaluation, 27, 163–184.

Harroch, R. 2000. Business Contracts Kit for Dummies. New York, NY: Hungry Minds.

Logan, M. 2004. Succession Planning Using Microsoft SharePoint. Retrieved on August 11, 2011 from http://www.knowinc.com/pdf/Succession_Planning_using_SharePointv2.pdf.

Loo R. 2002. Journaling: A Learning Tool for Project Management Training and Team Building. Project Management Journal. Philadelphia, PA: Project Management Institute.

Webster, B., Hare, C., & McLeod, J. 1999. Records management practices in small and medium-sized enterprises: A study in north-east England. Journal of Information Science, 25, 283–294.

Yusof, Z., & Chell, R. 2000. The records life cycle: An inadequate concept for technology-generated records. Information Development, 16, 135–141.

Suggested Reading: Digital Projects Advisory Group. 2008. Guidelines on File

Research Methods

Angrosino, M. 2007. Doing Ethnographic and Observational Research: The Sage Qualitative Research Kit. Thousand Oaks, CA: Sage Publications.

Banks, M. 2007. Using Visual Data in Qualitative Research: The Sage Qualitative Research Kit. Thousand Oaks, CA: Sage Publications.

Barbour, R. 2007. Doing Focus Groups: The Sage Qualitative Research Kit. Thousand Oaks, CA: Sage Publications.

Carmines, E. & Zeller, R. 1979. Reliability and Validity Assessment. Thousand Oaks, CA: Sage Publications.

Cresswell, J. & Clark, C. 2007. Mixed Methods Research. Thousand Oaks, CA: Sage Publications.

Flick, U. 2007. Managing Quality in Qualitative Research: The Sage Qualitative Research Kit. Thousand Oaks, CA: Sage Publications.

Flick, U. 2007. Designing Qualitative Research: The Sage Qualitative Research Kit. Thousand Oaks, CA: Sage Publications.

Gibbs, G. 2007. Analyzing Qualitative Data: The Sage Qualitative Research Kit. Thousand Oaks, CA: Sage Publications.

Golden-Biddle, K. & Locke, K. 2007. Composing Qualitative Research. Thousand Oaks, CA: Sage Publications.

Hunt, M. 1997. How Science Takes Stock: The Story of Meta-Analysis. New York: NY: Russell Sage Foundation.

Kvale, S. 2007. Doing Interviews: The Sage Qualitative Research Kit. Thousand Oaks, CA: Sage Publications.

Lewins, A. & Silver, C. 2007. Using Software in Qualitative Research: A Step-by-Step Guide. Los Angeles, CA: Sage Publications.

Lipsey, M, & Wilson, D. 2001. Practical Meta-analysis. Thousand Oaks, CA: Sage Publications.

Littell, J., Corcoran, J., & Pillai, V. 2008. Systematic Reviews and Meta-Analysis. New York, NY. Oxford University Press.

Marshall, J. & Mead, G. 2005. Action Research. Thousand Oaks, CA: Sage Publications.

McMillan, J. 2004. Educational Research: Fundamentals for the Consumer 4th Ed.Boston, MA: Pearson Education, Inc.

Miles, M. & Humberman, M. 1994. Qualitative Daa Analysis 2nd Ed. Thousand Oaks, CA: Sage Publications.

Rapley, T. 2007. Doing Conversation, Discourse and Document Analysis: The Sage Qualitative Research Kit. Thousand Oaks, CA: Sage Publications.

Straus, A. & Corbin, J. 1998. Basics of Qualitative Research: Techniques and Procedures for Developing Grounded Theory 2nd Ed. Thousand Oaks, CA: Sage Publications.

Wolf, F. 1986. Meta-Analysis: Quantitative Methods for Research Synthesis. Thousand Oaks, CA: Sage Publications.

Yin, R. 2008. Case Study Research: Design and Methods. Thousand Oaks, CA: Sage Publications.

Measurement Practices

Bogan, C. & English, M. 1994. Benchmarking for Best Practices: Winning Through Innovative Adaptation. New York, NY: McGraw Hill.

Camp, R. 1995. Business Process Benchmarking: Finding and Implementing Best Practices. Milwaukee, WI: ASQ Quality Press.

Cizek, G. and Bunch, M. 2007. Standard Setting: A Guide to Establisha dn Evaluating Performance Stands on Tests. Thousand Oaks, CA: Sage Publications.

Damelio, R. 1995. *The basics of benchmarking.* Portland, OR: Productivity Press.

Darton, M. & Clark, J. 1994. The Macmillan Dictionary of Measurement. New York, NY.

Ory, J. & Ryan, K. 1993. Tips for Improving Testing and Grading. Newbury Park, CA: Sage Publications

Kanji, G. 2006 100 Statistical Tests 3rd Ed. Thousand Oaks, CA: Sage Publications

Kurpius, S., Stafford, M. 2006. Testing and Measurement. Thousand Oaks, CA: Sage Publications

Morgan, S., Reicher, T., & Harrison, T. 2002. From Numbers to Words. Boston, MA: Pearons Education, Inc.

Salkind, N. 2006. Statistics for Peole Who Think They Hate Statistics. Thousand Oaks, CA: Sage Publications

Salkind, N. 2006. Tests & Measurement for Peole Who Think They Hate Tests & Measurement. Thousand Oaks, CA: Sage Publications

Straw, J. 2002. The 4-Dimensional Manager: DISC Strategies for Managing Different People in the Best Ways. San Francisco, CA: Inscape Publishing.

Upton, G. and Cook, I. 2002. Oxford Dictonary of Statistics. Oxford, NY: Oxford Univerity Press.

Watson, G. 1992. The Benchmarking Workbook: Adapting Best Practices for Performance Improvement. New York, NY: Productivity Press.

Supply Chain Management

APICS. 2009. Certified Supply Chain Professional CSCP Certification Training Guide. Chicago, IL: APICS.

APICS. 2011. APICS Supply Chain Manager Competency Model. Chicago, IL: APICS.

APICS 2009. Operations Management Body of Knowledge OMBOK. Chicago, IL: APICS.

APICS 2010. APICS Dictionary. Chicago, IL: APICS.

Arnold, T., Chapman, S., & Clive, L. 2001. Introduction To Materials Management. Upper Saddle River, NJ: Columbus, OH.

Arnold, T., Chapman, S., & Clive, L. 2001. Introduction To Materials Management Casebook. Upper Saddle River, NJ: Columbus, OH.

Brown, S. 2000. Customer Relationship Management: A Strategic Imperative in the World of e-Business. Etobicoke, Ontario: John Wiley and Sons.

Jain, C. & Malehorn, J. 2005. Practical Guide to Business Forecasting. New York, NY: Graceway Publishing.

Lee, Q. & Snyder, B. 2006. Value Stream & Process Mapping. Bellingham, WA: Enna Products Corporation.

Sheffi, Y. 2005. The Resilient Enterprise: Overcoming Vulnerability for Competitive Advantage. Boston: MA: MIT

Visualization and Presentation

Damelio, R. 1996. The basics of process mapping. Portland, OR: Productivity Press.

Few, S. 2006. Information Dashboard Design: The Effective Visual Communication of Dat., Sebastopol. CA: O'Reilly Media.

Harris, R. 1999. Information Graphics A Comprehensive Illustrated Reference. New York: Oxford University Press.

Kodukula, P. Meyer-Miller, S. 2003. Speak with Power, Passion and Pizzazz! 222 Dynamite Tips to Dazzle Your Audience. Hats off Books.

Mills, H. 2007. Power Points!: How to Design and Deliver Presentations That Sizzle and Sell. New York, NY: Amacom.

Mucciolo, T. & Mucciolo, R. 2003. Purpose, Movement, Color: A Strategy for Effective Presentations. New York, NY: MediaNet.

Reynolds, G. 2008. PresentationZen: Simple Ideas on Presentation Design and Delivery. Berkeley, CA: Newriders.

Concluding Thoughts

"Faith believes a lesson is real."

"Hope is thinking the lesson can have value."

"Trust in a lesson usually involves a choice."

"Wishful thinking is that a lesson will make things better."

"There are lessons to learn all the time—some good, some bad, and some neither good or bad."

"Depending upon a lesson and trusting in a lesson are two different things."

"If it isn't documented, then it didn't happen."

Index

About the Author

Willis H. Thomas, PhD, PMP, CPT is a certified project manager and performance technologist who has been involved in organizational development and effectiveness across the pharmaceutical, telecommunications, information technology, and security industries.

Willis holds a PhD in Evaluation from Western Michigan University. The title of his dissertation while working at Pfizer was *"A Metaevaluation of Lessons Learned Repositories: Evaluative Knowledge Management Practices in Project Management."* In line with his extensive research, he has developed a comprehensive website on project management that focuses specifically on lessons learned at www.lessonslearned.info.

Willis earned his Master's of Science in Human Resource Management from National-Louis University. The title of his Master's thesis, while working at Xerox, was *"The Impact of Sales Integration on Interpersonal Communication Affecting Sales Representatives Job Satisfaction."* He began his career selling copiers to small businesses and transitioned to selling information systems. After leaving Xerox, he worked for Ameritech and sold telecommunications products and services. He later transitioned to human resources.

A unique and challenging opportunity presented itself with Brinks. It was a diverse training and project management role supporting human resources, sales, operations and marketing. After building many programs, he became a consultant for five years and developed marketing programs, websites, and training programs for companies, including GTE (Verizon), DNSE, JP Systems, ODS Networks, Citizens Communications, BT Office Products, Itronix, and many others.

His move to Michigan was prompted by a desire to enter the pharmaceutical industry due to significant changes in the telecommunications industry and complete a PhD in Evaluation at Western Michigan University. Joining Pfizer offered an unique opportunity to do both.

Willis' Bachelor's of Arts degree is from the University of Wisconsin-Madison. While there, he served as the editor for *Datelines* Student Newspaper and editor for West African Monsoons from Geocentric Satellites (Space Science and Engineering Department).

He has earned and maintains a Project Management Professional (PMP) certification from the Project Management Institute (PMI) and a Certified Performance Technologist (CPT) certification from the International Society of Performance Improvement (ISPI). In the area of project management, he has taught more than a dozen courses worldwide for the International Institute for Learning.

Willis has also served as an adjunct professor, instructor, and guest speaker.